The Anatomy of a Flying Saucer

Detailed Scientific Explanation of How UFOs Work

by

John Mike

ISBN 9781463598068
ISBN 10-1463598068

Table of Contents

The Anatomy of a Flying Saucer

Origins

Politics

Miscellany

FOREWORD

This book owes its existence entirely to work of one man, Dr. J. Allen Hynek, who is usually referred to as the Father of UFOlogy. Dr. Hynek was for many years the spokesperson and front man for the United States Air Forces program to collect information from the public about UFOs, eventually called ***Project Blue Book***. Dr. Hynek was a real scientist, at one time the Chairman of the Astrophysics Department at Northwestern University.

I became interested in the subject three years ago when I happened to read a lovely little book called ***Night Siege*** co-authored by Dr. Hynek. In his most scientific book, ***The UFO Experience***, he presents detailed case accounts and describes his disillusionment with the Air Force. He tried to get the Air Force to do real scientific studies of the UFO phenomenon but he was unsuccessful. He then founded the Center for UFO Studies (CUFOS) to try to promote the research himself, but without government funding. Part of CUFOS work has been to maintain and build the ***UFOCAT*** searchable database of eyewitness reports original started by Dr. David Saunders at the University of Colorado. This database now contains some 209,550 reports. It is maintained by Dr. Donald Johnson, Dr. Mark Rodeghier, the Scientific Director at CUFOS, and others.

Besides the two books mentioned by Dr. Hynek, the present book is essentially based on the eye witness reports in the ***UFOCAT***. The database is in the form of a Microsoft Access file, and one can write queries to project out any portion of the vast amount of material available in the database. And that is what I did for this book. If I was interested in some aspect of the subject, I would project out the relevant reports, usually in the hundreds, to get some feel for the subject. But more interestingly, when I came to some interim conclusion as to how things might be, I would turn to the database, look, and surprisingly found the matter actually reported in eye witness accounts. In terms of physics, one might say I was forming hypotheses, and found verification in "experiment", the eye witness reports.

In the ***UFO Experience*** Dr. Hynek was mainly concerned with the vexing problem of the believability of UFO eye witness reports in general. That problem may be unsolvable. For this book I chose a road that simplified matters considerably. I concentrated on the reports from the sightings of two major waves, of the Hudson Valley 1983-1985 and of Belgium 1990-1992.

There were two major "filters" that mitigated the believability problem. First, the reports in both cases were taken from the eyewitnesses by experienced investigators. They lead the eyewitnesses in the reporting by asking questions. This had the effect, similar to a court of law, where the lawyer ask specific questions to elicit testimony and just does not let the witness ramble. Secondly, both of the waves were of the same type UFOs, Big Black Triangles, and the occupants of these craft did not appear (except possible for one case.) Therefore the whole sociological problem of aliens and abductions simply did not enter. Essentially the reports were of the craft and the crafts behavior, and from these reports a great deal could be deduced.

This book is entirely based on these eyewitness reports. There are no ad hoc theories imposed artificially on the material. The key insight of the book, that UFOs possess negative mass, is dictated directly by the reports. It became clear eventually that UFOs could not do what they do, unless they had negative mass. For a UFO to lift a car off the road a strong negative vector potential A(-) is needed, and that can only come from negative mass. Bondi showed in 1957 the General Relativity does not forbid negative mass. We on Earth have not yet found negative mass, but then UFOs are from other times (or possibly places).

In reading through ***The UFO Experience*** one finds many aspects of UFOs that Dr. Hynek wanted to understand. Indeed, that is why he founded CUFOS. And it now through the work of CUFOS, especially the ***UFOCAT***, that we have found answers to many of Dr. Hynek's questions. One might say that this book is a direct consequence of Dr. Hynek's work and that of CUFOS.

The numbers cited from ***UFOCAT*** are raw data from the queries. I did not read each report. They contain duplicates, misreads, and some fanciful material which is hard to believe. I did sample analyses of the reports. It would be a conservative estimate that half of the reports are substantive and on point. Since the numbers cited are in the hundreds, even the half of the numbers cited yields convincing data.

A note about the illustrations. Unfortunately they are only suggestive, and not accurate. Software to generate the vector potential curves accurately was not available. They are mostly analogies from electricity and magnetism

The book can be read profitably without being overly concerned about the equations and calculations. The text and the diagrams should give ample explanation. I included the calculations to show that the UFO explanations could be done in mathematical detail. Only chapters 13 and 19 are written explicitly for physicists.

Overview

1 Introduction

The conventional wisdom, promoted by the government and echoed by the media, is that UFOs are mysterious objects which by definition are unknowable. Therefore by that logic anyone attempting to explain UFOs must be a charlatan perpetrating a hoax employing "junk phyics". That may not be so.

This book explains how UFOs work. It solves essential mysteries of UFO behavior that have been well-documented and unsolved for over fifty years.

How is this possible? Very simple. The Center of UFO Studies (CUFOS) has for 30 years been accumulating a searchable database of UFO eyewitness reports. To date they number 209,000. To my knowledge no one has taken a serious look at these reports. I now have.

Just out of curiosity, I became interested in UFOs three years ago.
I new absolutely nothing about the subject. Everything I have learned has come from two books by J. Allen Hynek (who founded CUFOS) and the ***UFOCAT*** database.

All the conclusions in the book come directly from interpreting the eyewitness reports. The basic conclusions were an inevitable consequence of applying standard physics principles to the reports.

The fundamental revelation, that UFOs possess gravitationally negative mass, is an inescapable consequence of the fact that UFOs sometimes "levitate" people and cars.

Gravitationally negative mass, when spun at high speed, allows UFOs to negate inertia. This fact suddenly throws the whole field of UFO behavior open. Now essentially all the activities of UFO can be understood. They cannot only be qualitatively understood, but they can be calculated and predicted in detail. We do the calculations in the book and provide diagrams.

We not only explain how UFOs kill inertia, but how they float in gravity, why they do zig-zag motions instead of smooth turns, why they sometimes exhibit "falling leaf" motion, how they propel themselves, and why they sometimes levitate objects. We explain the famous EM effects of car engines and headlights dying near UFOs, and why they do not create sonic booms at supersonic speed.

The laws of physics in the universe appear to be uniform. As astronomers peer back to the beginning of time near the Big Bang, the universe seems the same. There are exotic objects like **Black Holes** and Dark Matter. These objects do not violate physics principles themselves, but they create strange environments, like matter disappearing behind the **Black Hole's** event horizon.

UFOs however seem to violate the laws of **Conservation of Momentum**, **Newton's Second Law of Motion**, and **Second Law of Thermodynamics**. But if we introduce another exotic object, negative mass, which like **Black Holes** in itself is NOT contrary to physics, the environment that negative matter creates, namely inertia free regions, removes the apparent violations of physical laws, and the laws of the universe are uniform again.

It is possible that UFOs have technologies that are far beyond us. Then we won't know what they are till they tell us. But what I find interesting is, that with the mere introduction of the concept of negative mass, a totally consistent picture of UFOs can be constructed based only on technologies known to us now. And in this picture UFOs obey all known laws of physics. What is even more interesting is that this picture is so simple and consistent with eyewitness accounts.

In retrospect, it appears that most of what is in this book has probably been known for over fifty years (see chapter on "The Obvious"), except nobody bothered to tell the rest of us (see "The Separation of UFO and State").

This book only peels back the first layer of mysteries, the physical behavior of UFOs. As you expect, it exposes a deeper layer of mysteries.

UFOs do not come from outer space. They appear, visually and on radar, only at lower altitudes. Hundreds of eye witness in the ***UFOCAT*** report how UFOs "shrink" out of existence. It has been known for at least one hundred years that the "shrinking" appearance is actually a hallmark of an object entering another dimension. (See "Escaping to Anther Dimension"). Other facts about UFOs reinforce this conclusion.

And there is evidence based on logic that UFOs do not come from a far away planet, like ET. The greatest living expert on UFOs, Dr. Jacques Vallee, noted years ago that there are just too many of them, probably over a million visits during the last century alone. UFOs are terrestrial. (See "UFOs R Us.")

After dealing with the subject for three years, I have come to the conclusion that UFOs are messengers of deeper mysteries of our physical world. By being able to enter other dimensions, they point to the fact our Einstein universe of gravity and galaxies is only a part of something larger.

2 Solve a Puzzle and Illuminate a World

In the mysterious world of UFOs there is a somewhat rare but puzzling phenomenon called "levitation". The ***UFOCAT*** database of 209,550 eye witness UFO reports, about 300 mention people being lifted up, of car steering lost, and in about 100 of these, cars is lifted off the road by a passing low flying saucer. 24 of these reports are simple, credible statements (see Chapter on Levitation). The reports are from four continents over a period of 40 years.

Now this is puzzling. Because the lifting forces involved can not be electromagnetic. You have seen cranes with large magnets pick up scrap metal in a junk yard. Magnetic forces are of extremely short range. The magnet in the yard actually touches the metal. The flying saucer that lifts car usually passes at a distance roughly of 20 feet from the car, too far for a magnetic field. It is not an electric field, because electric fields exert their effects on metal objects, not tissue. The reports indicate that the witnesses felt THEMSELVES to be levitated. Also high electric fields, as before a lighting strike, make people's hair stand on end, and this was not reported. So it must be some other force.

The only other force is gravity. Gravity effects all matter equally, whether it be metal or tissue. The interesting thing is that it is not a stationary flying saucer which did the lifting. The lifting occurred only when the saucer was moving. What that means is the gravitation involved is not Newtonian gravitation, like the gravitational field of the Earth. The source of this gravitational field is different.

The force of gravity is **F = mass x gravitational field**. According to Einstein the gravitational field **g** (usually called **E** by others) is made up of two parts, one is Newtonian gravitation

$$\mathbf{g}_N = -\nabla\varphi$$

where φ is scalar gravitational potential given by its mass divided the distance r from the body

$$\varphi = \mathbf{mass/r.}$$

The other term is the inertial term, the source of inertia and Newton's Second Law of Motion, involves the gravitational vector potential, **A**. But it is not **A**, but the time rate of change of **A**, **∂A/∂t**, which is a gravitational field

$$\mathbf{g_I = - \partial A/\partial t.}$$

Therefore the force of inertia is

$$\mathbf{F_I = mGg_I}$$
$$\mathbf{= - mG\partial A/\partial t}$$

where G is the gravitational constant.

So if it is not Newtonian gravitation which lifts the car, then it must inertial gravitation. What this implies is the lifting flying saucer is source of a strong vector potential **A**. As the saucer moves, the movement of the saucer causes a change in the vector potential of the saucer as felt by the car, and this movement creates a **∂A(-)**/∂t, a gravitational field which picks up the car.

Is it possible to make gravitational vector potential at all? Well, the answer lies in the definition of the vector potential. We here on Earth are incapable of creating a vector potential. As a matter of fact, we are unable to make gravitational vector potential at all. The US Government gave out secret contracts for twenty years for "anti-gravity" research. The research failed to produce any results, and the programs were cancelled in 1975. The only vector potential we know is the one involved in inertia, and that acts only on the accelerated body itself.

So our first conclusion:

UFOs can make gravitational vector potential, we can't.

The vector potential is defined by

$$\mathbf{A} = \int \rho \mathbf{v}/\mathbf{r}\, d\mathbf{V}$$

where ρ is the mass density, **v** the velocity, **r** the distance to the mass, and dV the element of volume being integrated over.

If we want to make measurable vector potential **A**, we must use either lots or very dense mass, or move it very quickly. But how do you move the mass fast, say in a UFO, if it is stationary? The answer is: you spin the mass at a high speed.

Here we run into sever limitations. First we can not make any material much more dense than lead at about 11 gm/cc. Transuramic artificial elements can be made which are slightly denser, but they disintegrate immediately. If we are going to put the mass into a flying saucer, we can't put a whole lot since a typical small saucer is only 30 feet in diameter. That is about the size of an F-22. An F-22 weighs 38 tons loaded, so, say, we put about 40 tons of mass into our saucer.

What shape would the mass be in to spin it? Well, it could be a disk or a ring. When we actually do the calculations it turns out that it has to be a ring. The only other variable we can use is the rotation speed of the matter.

So how fast do we have to spin it? If you spin a 30 foot 40 ton ring, it turns out you have to spin it at 5 million revolutions per second. The velocity of the rim at this point is half speed of light.

And then you are going to tell me that's crazy. When you spin an object there is centrifugal force that tries to tear the spinning mass apart. The Iranians had trouble spinning 6 inch diameter centrifuges at 60,000 rpm, or 1,000 revolutions per second, and their centrifuges disintegrated. So how are you going to spin something much larger 5,000 times as fast?

And that is the problem. We can't spin anything really fast because of inertia. Because as you try to make an object go around in a circle, each piece of the object is changing direction continuously. And this change in direction is an acceleration, and by Newton's Second Law this acceleration needs a force to accomplish it. The internal structural strength of the material tries to provide this centripetal force. But at some point the inertia and Newton's Second law overcome the strength of the material and it just flies apart. Therefore it is inherently impossible to spin, matter that we know, at speeds required to produce a substantial gravitational vector potential.

When you spin a body, it becomes what is called a gravitational dipole. In electromagnetics the dipole is a current loop where the electrons flow through a circular wire. But gravitational "charge" is the mass itself. And instead of the charged electrons flowing through the wire, in gravity we must swing the mass (or gravitational charge) itself. **A** spinning mass ring is analogous to the wire loop. The gravitational dipole then produces the vector potential **A**.

We know a gravitational vector potential from the UFO lifts the car. But how? There has to be an answer.

The only experience we have with the vector potential **A** is in the case of inertia. As a particle moves in our universe which is has a universal background scalar potential **Φ** with a velocity **v**, the motion creates a vector potential **A** (see **ARTICLE**)

$$\mathbf{A} = \mathbf{v\Phi/c^2}$$

If we take the time derivative of this equation, we ge

$$\mathbf{\partial A(-)/\partial t = a\Phi/c^2}$$

where $\mathbf{a} = \partial \mathbf{v}/\partial \mathbf{t}$ is an acceleration.

The force of inertia is

$$\mathbf{F_I = -\ m_I G a \Phi/c^2}$$
$$\mathbf{= -\ m_I G \partial A(-)/\partial t}$$

Note the force of inertia F is OPPOSITE to the acceleration and opposite to $\mathbf{\partial A(-)/\partial t}$.

When the car is sitting in the saucer's vector potential field and the saucer is stationary nothing happens. As the saucer moves, the **A** changes and there is a $\mathbf{\partial A(-)/\partial t}$ at the car. The car feels a gravitational force. But the force is in the OPPOSITE direction from the saucer. What this means is that the force from normal **$\partial A/\partial t$** is actually AWAY from the saucer. It actually pushes the car INTO the ground, not lift it up! This is NOT what is observed.

What would LIFT the car?

> **There is only one possibility. The SIGN of A has to change.**

How do we change the sign of **$A = \int \rho v/r\, dV$** ? We cannot change **v** or **r**. The only sign we can change is that of the ρ, the mass density: **$-A(-) = \int(-\rho)v/r\, dV$**

Then **$F = + mG\partial A(-)/\partial t$** is positive. It LIFTS the car.

What does that mean? What it means is that the mass in the saucer has to have a GRAVITATIONALLY NEGATIVE CHARGE, which then produces a NEGATIVE vector potential **A(-).**

All matter that we know in the universe has a positive gravitational charge. Can negatively charged mass can exist? Bondi showed in 1957 that gravitationally negative mass does not violate General Relativity. No one has ever refuted this conclusion. We know WE don't have gravitationally negative mass on Earth, not even in the laboratory. People have looked for it astronomically in the universe and have not found any.

> **What it DOES mean is that UFOs appear to have gravitationally negative mass.**

Let us now pause. Are we dealing here with some algebraic trick whereby we can make the phenomenon of a flying saucer lifting a car appear plausible by somehow changing a sign, - or does it imply more?

What happens is that suddenly everything begins to fall into place. If we allow that UFOs have negative mass, then ALL their mysterious behavior of UFOs becomes immediately transparent.

> **It is as if in a darkened stadium the lights were suddenly turned on that the entire field illuminated, so all action on the field can be seen and understood.**

Because now we can begin to understand what is happening. First of all it negative mass means is that the negative mass CAN be spun at great speeds.

The reason is that if a NEGATIVE vector potential, **A(-),** is produced, and this negative potential repulses the positive **A(+)** of the universe. Why? Because negative and positive gravitational fields repel each other. One excludes exclude the other. The boundary between them has to be zero .

If it is a negative gravitational vector potential which lifts the car, then the region in which the saucer and the car are has a negative vector potential **A(-)**. It can not have a positive vector potential **A(+).**

But by **Mach's Principle** it is the "distant stars", or the universal scalar potential **Φ** which is responsible for the positive vector potential **A(+).** and it is a change in **A(+)** which is responsible for inertia. Einstein implied this in his 1952 book ***The Meaning of Relativity.*** Sciama showed it in a calculation in 1957. A much more refined, accurate, and relativistic calculation was done in 2007 be Martin, et. al. The paper was called "Inertia as Gravitation." It involved integrating over the entire mass of the universe, including uncertain things like dark energy. The calculation confirmed **Mach's Principle** finding within 10%. Therefore that the vector potential **A(+)** causes inertia is well established.

If the positive **A(+)** of the universe creates inertia, and it is excluded from the saucer-car region, then by **Mach's Principle** (or if you wish, its corollary) there is no inertia in that region.

The negative mass can be spun at high speed because the mass has no inertia, no centrifugal force is created by spinning it, and no centripetal force is required to hold it together.

.

And THIS is the source of the revelation, of the stadium lights. Because if there is NO inertia around a UFO, then ALL the weird things that happen around UFOs can now be readily explained. (See the chapter "The UFO Magical Mystery Tour".)

For sixty years witnesses have reported that car engines and headlights die near flying saucers. These are the famous EM effects. They featured prominently in the movie ***"Close Encounters Of the Third Kind"***. Now these can be understood.

A window alarm attached to the window frame has a small magnet near it attached to the sash. If the sash is opened, the magnet is moved away and the alarm sounds. This is because of **the Hall Effect**. The window alarm uses a semiconductor where the structure of the semiconductor reduces the effective mass of the electrons to a few percent of their normal value. Since the momentum of the electron is, $\mathbf{m_i v}$, inertial mass times the velocity, its momentum is now small. The charge of electron remains unchanged, so a magnetic field can interact with the charge to easily bend the path of the electron with weakened momentum. The small magnet can deflect the current by the **Hall Effect**. The magnet bends the path of the electrons in the semiconductor. The electrons are diverted. They do not reach the anode. Remove the magnet and the current flows and the alarm sounds.

If there is no inertia in the region around a UFO, then the inertial mass of the electrons effectively approaches zero. The electron momentum is now even smaller. Any stray magnetic field, including that produced by the cars own wiring, can deflect the current by the **Hall Effect**. A 10 amp current in the car's wiring produces a 30 gauss magnetic field, and this stops current flowing, just as the magnet of the window alarm. Car engines and headlights die. When the flying saucer leaves, inertia comes back, the electron momentum returns to it normal value, car engines and headlights come back on.

When traveling at supersonic speeds UFOs do not create sonic booms. Why? Because the air molecules have almost no inertial mass in the inertial free zone around the UFO . Therefore the air molecules are easily pushed out of the way. They do not pile up in front of the craft to create a shock wave which normally causes a sonic boom.

The large vector potential produced by UFOs is not there to kill engines and headlights and prevent sonic booms on Earth. It is there to kill INERTIA in the UFO. It is the lack of inertia in a UFOs which explains why they can do those fantastic feats of acceleration, reportedly up to **35**g, without making a sound (and fatally slamming their occupants around). They can do that ONLY because they have almost no inertia. UFOs use their negative mass to produce all that negative vector potential **A(-)** PRECISELY to cancel inertia.

We calculate later in the book (chapter on "UFO Anti-Gravity Propulsion") that a UFO with a mass of hundreds of tons has an inertial mass of about 1 kg, or **one one millionth** it would have without the negative mass. UFOs and their occupants are essentially free of the inertia felt by everything else in the universe. The UFOs can act more like light beams then massive craft.

These mysteries have surrounded UFOs for decades. If we now allow that they have negative mass, these mysteries can not only be qualitatively explained, they can be calculated and predicted in precise detail using ordinary physics.

Therefore solving the puzzle of how a small passing flying saucer can lift a car, suddenly throws a glaring light of illumination on the whole field of UFOs. And that is what we will explore in this book.

3 The UFO Magical Mystery Tour

THE MAGIC: Ten Magical Things That UFOs Do That Are Not Possible In the Universe As We Know It.

1. Make high speed zig-zag motions (act like light beams, not massive craft)
2. Great accelerations without sound (no sign of energy or thrust use for acceleration)
3. Float in gravity slowly in silence (no energy or effort to stay aloft)
4. Fly faster than sound without a sonic boom.
5. Propel themselves silently without interacting with the environment (no visible thrust, turbulence, heat, light)
6. Exhibit Pendulum or "Falling Leaf" Motion
7. Cause car engines and headlights to die (there is no force on earth that can turn a car off and on at a distance with no after effect)
8. Cause temporary paralysis (paralysis that comes and goes with presence of UFO)
9. "Levitate" people and cause steering loss (no electric or magnetic forces can lift human tissue or cause steering loss)
10. Lift cars off the ground. (No known force can lift a car from 20' away)

THE MYSTERY SOLVED: One Simple Unremarkable Fact Explains All.

UFOs have negative mass.
Herman Bondi showed in 1957 that Negative Mass does not violate General Relativity
This has never been challenged.
Physicists have looked for and theorized about it.
Its discovery would not be remarkable.

In a Nutshell:

The movement of UFOs violates the **Law of Conservation of Momentum**, **Newton's Second Law of Motion**, and the **Second Law of Thermodynamics**. By the **Law of Conservation of Momentum** the well-known zig-zag motion of UFOs is not possible for massive objects. Tremendous accelerations of 22g and 35g reported by radar for UFOs are only possible by exertion of great force according to **Newton's Second Law of Motion**, requiring the expenditure of huge amounts of energy. This energy has to be dispersed by the **Second Law of Theremodynamics**. No energy dispersal is seen. If one introduces the notion of gravitationally negative mass, all the above violations of physical laws are resolved.

UFOs Have Negative Mass, We Don't

Positive and negative gravitational fields repel each other.
With negative mass UFOs can create a space that has only negative gravitational fields.
The effects of positive gravitational fields are excluded.
Our universe has only positive gravitational fields.
The effect of our universe is excluded from the negative mass regions.
UFOs create gravitationally negative spaces, mini universes that are holes in our universe.
The negative gravitational space is a small bubble around the UFO.
The bubble travels with the UFO.
The radius of the bubble is inversely proportional to the square root of the UFOs speed:
(see **ARTICLE**)

r is proportional to 1/√v

At high speed the bubble is small. When the UFO slows down, **v** approaches zero, the bubble expands, to thousand or more feet in diameter.
Being gravitation, the bubble is not visible, but its

effects can be felt.

When the UFO is at low altitude, the negative gravity bubble breaks the ground plane. People and object now find themselves inside the bubble in an inertia free region.

The bubble around the UFO is very real.

Weird things happen around the UFO inside the bubble.

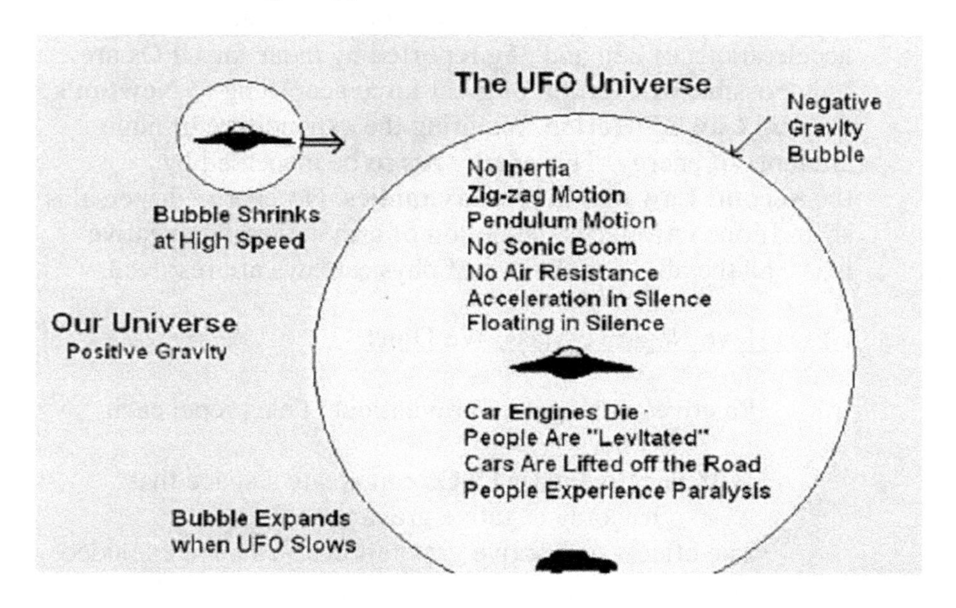

What happens inside this region has been reported for at least 60 years. Thousands of eye witnesses report failure of car engines and headlights, levitation and the loss of steering control.

The effects were accurately documented in the Spielberg 1972 film "***Close Encounters of the Third Kind***". The two greatest experts on UFOs,

Dr. J. Allen Hynek and Dr. Jaques Vallee were consultants on the film.

The weird things can essentially all be explained by the lack of inertia in the bubble which otherwise can not be readily felt, and the powerful negative gravitation field of the UFO

which, as the UFO
moves, causes levitation
All the effects as reported by eye witnesses can
now not only be explained, but calculated in
detail.
With this adjustment, all of the UFOs actions and
effects become understandable.

How UFOs Work

UFOs have equal amounts of positive and negative
mass. The gravitational forces on these cancel
each other out. UFOs float.
The gravitational field **E** is composed of two parts,
based on the potentials **φ** and **A**:

$$\mathbf{E} = -\nabla\varphi - \frac{\partial \mathbf{A}}{\partial t}$$

$\nabla\varphi$ is Newtonian gravitation, and $\partial\mathbf{A}/\partial t$ is
the inertial_term
The vector potential A is created when on object
moves in the universal background scalar
potential **Φ** that permeates the universe.
Acceleration of the object creates a $\partial\mathbf{A}/\partial t$ which
is a gravitational field, and $m\partial\mathbf{A}/\partial t$ the force
of inertia.
The negative mass in a UFO is in the form of a ring
(frisbee) that is spun at very high speed.
This creates a strong negative vector potential,
A(-).
The positive vector potential from the universe,
A(+), has a sign opposite to that of **A**(-).
When the two are added, if **A(-)** is larger,
the vector potential of the universe **A(+)** is
overcome. It does not reach the UFO.
The $\partial\mathbf{A}(+)/\partial t$ that creates inertial force
does not reach the UFO. The UFO is
free of inertia.
Because the UFO is free of inertia, it has essentially
no inertial mass.

It has a very small inertial mass, about **one millionth normal** or about 1 kg. This comes from the negative scalar potential of the ring itself.

If the spinning negative ring is shifted slightly, the positive mass structure of the UFO feels a ∂**A(-)/**∂**t.** This is a gravitational field. This propels the UFO (easily, because it only has the inertial mass of 1 kg.)

Why the Ten Things UFOs Do That Are Not Possible In Our Universe Are Possible (Normal) In Their Negative Gravitational Universe Or Bubble.

Because their inertial mass is about 1 kg, they can act as if they were essentially massless, like light beams, make zig-zag motions.

Since a 1 kg mass is easy to propel, they use very little energy, not enough to be detectable. Since **F = mass x acceler**ation, great accelerations are possible with little force.

Because they have equal amounts of positive and negative mass, they need no energy to stay aloft.

In the bubble there is (essentially) no inertia. Air molecules also are basically inertially massless. Consequently the molecules are easy to push out of the way. They do not pile up to cause a shock wave. Therefore no sonic boom (a shock wave) at high speed. There is little air resistance.

In the bubble where there is no inertia, electron inertial mass and momentum essentially vanish. The path of electrons can easily be bent **(Hall Effect**). Stray magnetic fields from the cars own wiring deflect electric current. Because the current is deflected, it stops flowing. headlights and engines die.

Nerve conduction in limb axons 3 to 4 feet long is also electrical. **Hall Effect** stops conduction. Result is

temporary paralysis.

When a UFO makes a close pass, people and cars are affected by the strong negative vector potential **A(-)** from the spinning ring. The movement of the UFO creates a **∂A(-)/∂t**, a change in the vector potential, which is a gravitational field. This is a force is felt as "levitation" by those effected. The force lifts the car's weight from the tires, they loose traction, causing steering loss. If the UFO passes close enough (20 feet or so) this gravitational force can actually lift the car off the ground.

The Explanation Of UFO Behavior Is Surprisingly Simple.

There are no "Theories" involved. The explanations are based entirely on eyewitness reports.

No new physics is needed. All that is need is the application of established principles of physics to the eyewitness reports.

Standard physics is applicable to negative mass.

Negative mass is not new physics. The question of Negative mass is just a question of empirical fact, i.e. whether it exists.

The implications of negative mass have not yet been thought through.

It turns out that the world of negative gravity is different from the world of positive gravity we are used to.

How Negative Gravity Is Different

In the micro negative universe which is the UFOs bubble, gravity behaves more like electromagnetism than the gravity we know from our gravitationally positive world.

All matter in our universe has a gravitationally positive charge.

Each body produces a scalar potential φ
All the **φs** of the universe accumulate into a gigantic Scalar background potential **Φ.**
Φ permeates the universe. **Φ** is like the glue that holds the universe together.
Sciama noted that **$G = c^2/\Phi$**. That the gravitational Constant is a function of **Φ**
It follows from general relativity that in our universe the vector potential **A** is given by (see "Dissertation on Mass")

$$\mathbf{A = G/c^2 \int\rho\ v/r\ dV}$$

But using Sciama's relation, **with $\Phi = 10^{29}$**

$$\mathbf{A = 1/\Phi \int\rho v/r\ dV}$$
$$\mathbf{= 10^{-29} \int\rho v/r\ dV}$$

The 10^{-29} factor is so small it makes all finite calculated fields vanish.
The only way **A** can be in any sense be measurable is if the integral is over the entire universe. This integral is the one responsible for inertia. Inertia is measurable.
In our universe, the presence of **Φ** cuts down all other fields to miniscule, immeasurable size. This is the implication of **Mach's Principle**.
But the negative gravitational fields in our UFO bubble exclude the effects of **Φ**. We do not multiply by **$1/\Phi$** because the effects of **Φ** are not present. The bubble is devoid of fields except those from the negative spinning ring. The effects of positive fields are excluded. There are no large accumulations of negative mass in the universe. People have looked and not found any.
Therefore there is no **Mach's Principle** for negative mass.
The situation is similar to electromagnetism. Positive and negative charges in matter cancel each other out. There is no accumulation of some universal electromagnetic field.
The formula for the electromagnetic vector potential is

$$\mathbf{A_{em} = (1) \int\rho_{em}\ v/r\ dV}$$

The integral is not multiplied by any field.
The coefficient in front of the integral is 1.
Because the negative gravitational field is isolated in the bubble, we will multiply by a constant **κ**, which has the value 1 (as in electromagnetism) and other units to make the dimension come out right.

$$\mathbf{A_g(-) = \kappa(=1) \int \rho_g\, v/r\, dV}$$

κ being 1, not 10^{-29}, the calculated negative vector field **A(-)** is now finite, not immeasurably small.

NOTE: Why according the Mach's Principle inertia is caused by the "distant stars"' lies in the mathematics of the integral for the scalar potential **Φ**. The force of inertia is given by

$$\mathbf{F_i = -\, mGE_i = -\, mG\frac{\partial A}{\partial t} = \frac{-mGa\Phi}{c^2}}$$

where **a** is the acceleration

F_i, the force of inertia, is directly proportional to **Φ**, the universal background scalar potential of the entire universe. To find **Φ** one must integrate over the entire mass of the universe. The formula is given by

$$\mathbf{\Phi = \int \rho/r\, dV.}$$

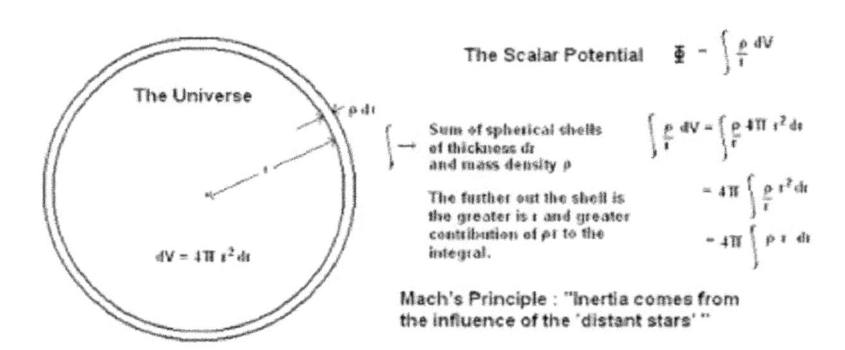

The integral is effectively a sum over thin shells of thickness

dr, each contributing **ρr**. Note that the further out you go with the shells, **r** becomes larger, with **ρ** essentially constant. Therefore the ring with the largest **r** or the furthest out shell contributes the most to the integral. The greatest contribution to **Φ**, and thus the force of inertia $\mathbf{F}_i$ will be the "distant stars".

THE PHYSICS

4 UFOs Have No Inertia

UFOs move in a zig-zag pattern. The ***UFOCAT*** listst 720 zig-zag reports. They cannot make continous smooth turns. No massive material object can do zig-zag motion. It is a violation of the **Law of Conservation of Momentum** and would require infinite energy. It is possible only for objects which have no inertia or inertial mass. Therefore the characteristic zig-zag motion of UFOs by itself proves that they are not subject to inertia and have (essentially) no inertial mass.

If you want to see UFO zig-zag motion for yourself, you can on the internet at: <http://video.google.com/videoplay?docid=-5146667346138302517#>

In November 1994, at the Nellis Test Range in Nevada where cameras and radar automatically track military fighters and their targets, a UFO was recorded and its motion photographed.

If you apply some simple physics, what you watch is not possible in the universe as we know it. Here is why.

Acceleration is a change in velocity

$$\mathbf{a} = \partial \mathbf{v} / \partial \mathbf{t}$$

and turning is also a change in velocity, and therefore an acceleration.

An Infinite Force is required for an Instantaneous Turn

acceleration = $\frac{\partial v}{\partial t}$

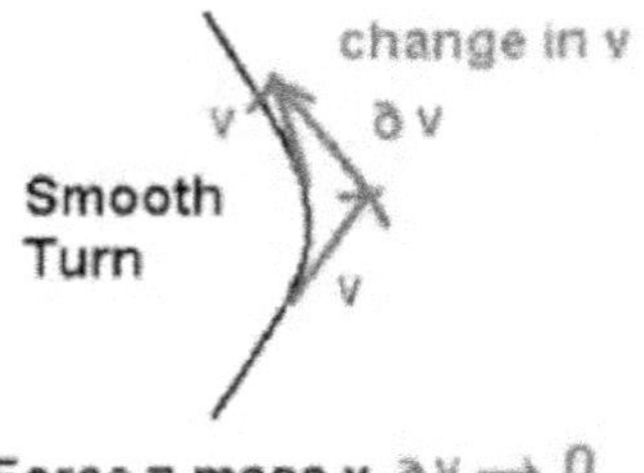

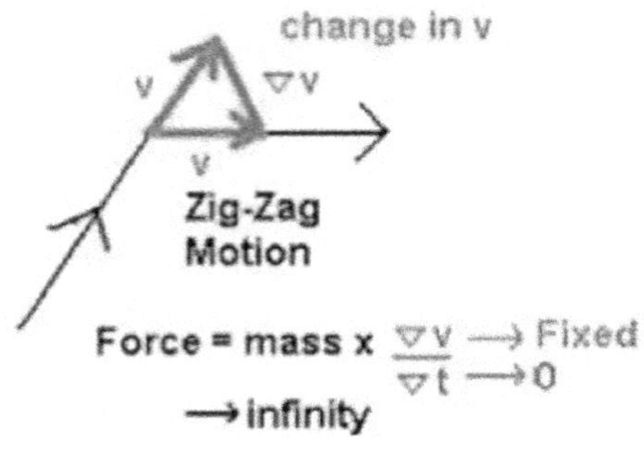

Force = mass x $\frac{\partial v \to 0}{\partial t \to 0}$
= finite

As $\partial t \to 0$, both ∂v and ∂t shrink, $\frac{\partial v}{\partial t}$ remains finite.

Force = mass x $\frac{\nabla v \to \text{Fixed}}{\nabla t \to 0}$
$\to$ infinity

As $\nabla t \to 0$, ∇v remains fixed. The $\frac{\nabla v}{\nabla t}$ ratio goes to infinity

We see that as an object turns smoothly, as we shrink the time interval ∂**t**, the change in velocity ∂**v** shrinks with it, and the ratio remains finite. On the other hand in a zig-zag turn, the change in velocity, **∇v** is discontinuous or fixed. In evaluating the acceleration, as we shrink the change in time **t**, **∇v** remains constant, and the ratio **∇v/∇**t actually goes to infinity. Since by Newton's Second Law

$$\mathbf{F = mass\ x\ (acceleration)}$$
$$\mathbf{F = mass\ x\ (\partial v/\partial t)} \text{ remains finite with } \mathbf{\partial t \to 0}$$

But

$$\mathbf{F = mass\ x\ (\nabla v/\nabla t) \to \infty} \text{ as } \mathbf{\nabla t \to 0.}$$

You need an infinite force to make an instantaneous change in direction. An infinite force is not possible. The only way that the equation balances is if the **mass = 0.** Then

$$\mathbf{F = 0 = (mass = 0)\ x\ (\nabla v/\nabla t).}$$

The mass in this case is the inertial mass, $\mathbf{m_i}$. One can distinguish between aspects of mass. Mass that measures the amount of matter in an object is **m**, the mass involved with gravitational attraction is gravitational mass, $\mathbf{m_g}$, and finally there is inertial mass, $\mathbf{m_i}$. We will discuss all these concepts at length in the chapter on "A Dissertation on Mass". Therefore,

UFOs have no inertia because their inertial mass with respect to moving in space is essentially zero.

In addition, zig-zag motion appears to violate the principle of **Conservation of Momentum**. It is illustrated by the well-known Newton's Cradle. The number of balls striking the row, equals the number of balls ejected

Newton's Cradle

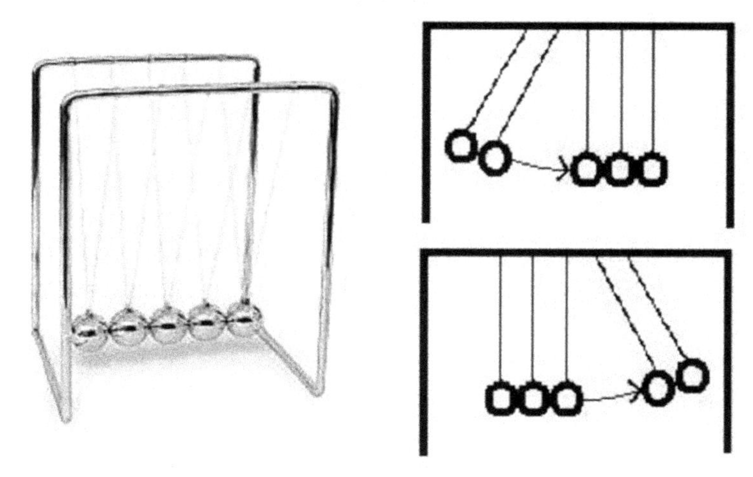

Demonstates Conservtion of Momentum

By that principle the total momentum of a system must remain unchanged under all interactions. In zig-zag motion the craft has momentum one direction and then suddenly in another.

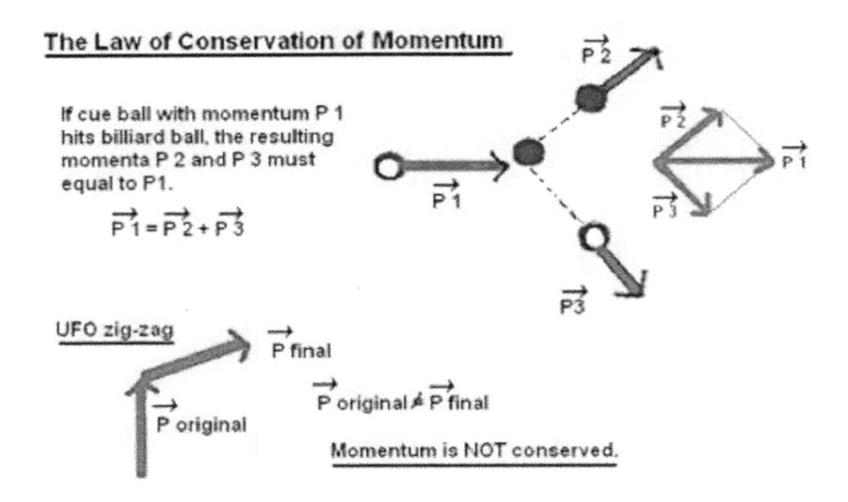

There appears to no transfer of momentum to the environment to balance this change. The UFOs motion is isolated and is the total system. Therefore momentum is apparently not conserved and merely changes in the zig-zag motion. The only way this can be is if the UFO has NO MOMENTUM in either direction, that is, the inertial mass is essentially zero.

There is an excellent analysis of the data of the Nellis object by Dr. Karl Ruada. It is on the internet at <http://www.roswellproof.com/Nellis_discussion.html>.

He states:
"The maximum acceleration was about 9g (9 times that of gravity). Such accelerations would have to be considered conventional, but found only on high performance craft that are jet- or rocket-propelled, yet the object exhibits no trail expected from such propulsion systems."

It is true that the Shuttle in its ascent, during its drive to reach orbit at one point goes through a 9g acceleration. Also fighter jets can pull 9g's in a turn or coming out of a dive. But that is not stop and go in level flight as we see in the Nellis data.

To better understand this, we should consider a more

maneuverable vehicle like the Air Force F-22. The F-22 Raptor has two engines, which with afterburners, produce 156 Kilo Newtons of thrust each, so the pair about 300 KN. The empty weight of the F-22 is 20 tons and 38 tons fully loaded.

If we take the weight as 30 tons and apply Newton's Second Law, then

$$F = ma$$

$$300\text{ KN} = 30\text{ tons x acceleration.}$$

$$\text{acceleration} = \frac{300\text{ KN}}{30\text{ tons}} = 10\text{ meters/sec/sec}$$

$$= 1\text{ g}$$

That is, the F-22 can do a 1g acceleration in straight horizontal flight on full afterburners. When the F-22 accelerates in level flight, the pilot will be pushed back in his seat by approximately
1 g, not 9g's as with the Nellis UFO. This agrees with the fact that the F-22 can basically fly straight up, just balancing gravity.

Say the Nellis object, which appeared to be 25' in diameter had a mass of 20 tons. Then its propulsive thrust or a 9.2g acceleration normally would have been

$$F = ma$$

$$= 20\text{ tons x }9.2\text{ x }10\text{ meters/sec/sec}$$

$$= 1840\text{ Kilo Newtons}$$

or nine times that of the F-22 or almost half of the Shuttle's 5400 Kilonewtons thrust. The problems is that it showed NO evidence of a "trail" from such a thrust.

What the analysis did not consider is the **Second Law of Thermodynamics**. The **Second Law** states that energy used has to be dispersed. The afterburners on the F-22 produce hot gas plumes and a giant roar, and rocket exhaust plumes of the Shuttle are hundreds of feet long, and can be heard up and down the Florida East Coast.

The Nellis object in its high acceleration maneuver not only did it not show any exhaust "trail", as the comment mentioned, it showed absolutely NO energy release of any kind. That means that no energy was expended in the maneuvers. That means there was no force or "thrust' involved.

In our universe, massive objects cannot make accelerations without energy or thrust. It is expected that when the UFO makes a 9.2g acceleration that requires half the thrust of the shuttle, that there be giant fireworks of exhaust plumes and roar. None of that happened with the Nellis object. Total silence. Nada.

The only way that could happen in our universe is that the UFO had essentially no inertial mass. That the UFO essentially did not experience inertia.

We have another documented example.

On March 30, 1990, two Belgian F-16s were scram led by the Belgian Air Force to intercept UFOs reported by SOBEPS, the Belgian society formed to investigate the UFO wave that Belgium was experiencing during 1990 and 1991.

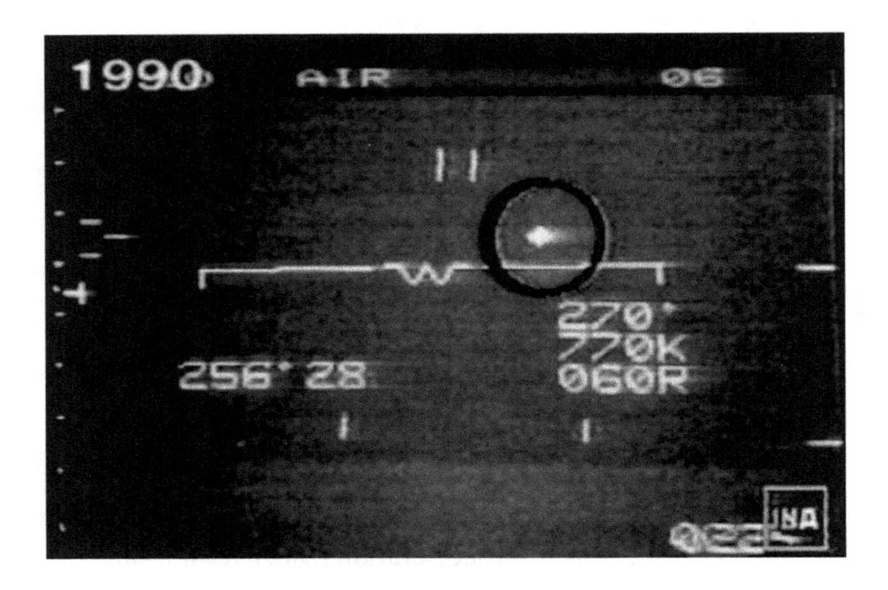

ehe UFO made a 22g acceleration, going from essentially zero They made repeated brief radar contacts with the objects. The radar was recorded on video tape. In one contact the UFO made a 22g acceleration, going from essentially zero to 500 mph in one second. We can see that from the excerpt of transcript below. (The entire official report by the Belgian Air Force is in the Appendix)

Seconds after lock-on	Heading (degrees)		Speed (knots)	Altitude (feet)
03	200		150	7000
04	sharp 200	acceleration	150	6000
05	turn 270	= **22g**	560	6000
06	270		560	6000

The Belgian UFOs were large Black Triangles. People often claimed they were 747s. A 747 weighs 200 tons empty, 387 tons full If we try to mach the pattern of their lights by superimposing their images, we see that while much of the UFO area is solid

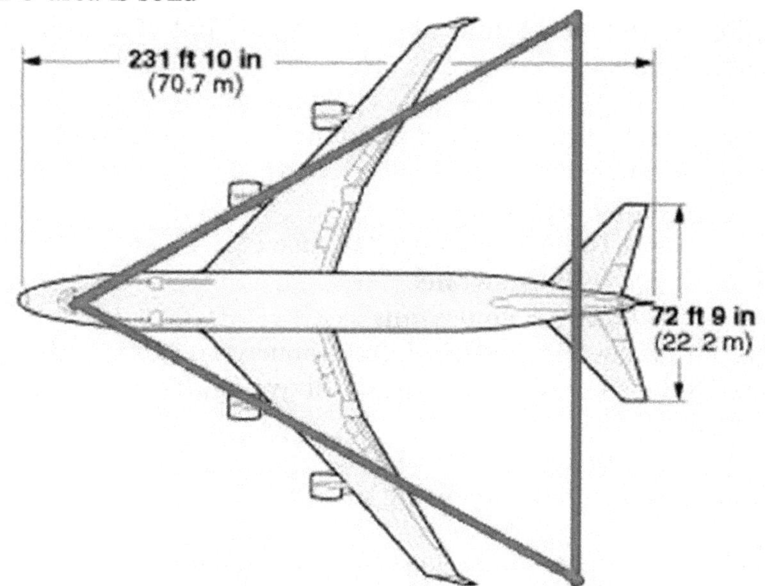

metal, most of the 747 is air behind the wings. A conservative estimate would be for the UFO to be 400 tons, but 800 tons is more likely.

What kind of force would be needed to accelerate the 747 from zero to 500 mph in one second? We get an idea by calculating the amount of energy involved. If the kinetic energy initially was small, we can calculate the final kinetic energy of the plane at 500 mph.

K.E. = 1/2 mass x $(\text{velocity})^2$

The mass is 400 tons = 400,000 kg

The velocity is 500 mph = 220 m/s

K.E. = 1/2 x 400,000 kg x $(220 \text{ m/s})^2$

$= (1/2)(4 \times 10^5)(2.2 \times 10^2)^2$

$= 9.68 \times 10^9$ joules

the amount of kinetic energy of the 747 sized object at 500 mph. The amount of energy in one KILOTON of high explosive,
4.18×10^9 joules.

Therefore the energy that would be required to make a 747 go from zero to 500 mph in one second would be equivalent to the **explosion of two kilotons of explosive, or that of a small atomic bomb**.

In terms of the force required to do the 22g acceleration

F = ma

= 400 tons x 22g

$= 4(10)^5$ kilograms x 220 m/sec^2

$= 8.8(10)^7$ newtons

= 88,000 Kilonewtons

This is 16 times the shuttle's 5,400 Kilonewton's thrust. If the mass Of the UFO was 800 tons, the above figures would be doubled.

We are faced with the following situation:

The radars in F-16s are designed to track and evaluate enemy craft. They detected and recorded on tape the radar contact with the UFO. This is what they are designed to do. They recorded a 22g acceleration by the UFO. The tape has been checked and the accuracy data has been verified. This is not swamp gas. This is not some hallucination by feeble minded eyewitnesses. This is hard data. It is about as close to a controlled experiment as you are going to get. The entire official report by the Belgian Air Force can be read in the Appendix A.

According to the **Second Law of Thermodynamics** energy used has to be dispersed. There are no exceptions. You have heard the rumble of a jet liner taking off, the fire, smoke, and roar of a top fuel dragster, and the magnificent spectacle of the shuttle taking off with the rocket plumes hundreds of feet long.

The force required to do the UFO acceleration is at least 88,000 kilonewtons. That is 16 times that of the shuttle. Therefore to do the 22g acceleration the UFO would have to create a rocket plume 16 times as large as the shuttle with 16 times the sound.

But it did not. It made no sound at all. It did the 22g maneuver in total silence. Also nothing was visible. The F-16s never made visual contact with the UFO. People on the ground saw the UFO but heard and saw nothing else.

Therefore the conclusion is this: By the **Second Law of Thermodynamics** energy used has to be dispersed. None was seen from the UFO. Therefore the energy used by the UFO to make the 22g acceleration was so small that it's dispersal could not be seen.

Later we will show that probable energy use of up to about 50,000 joules may not be apparent. What kind of an object can be accelerated to **22g**, or even **35g** (1), by 50,000 joules? We will calculate later that it is about 1 kilogram, or that the inertial mass, $\mathbf{m_i}$, is about **one millionth** of the actual mass, **m**.

If you believe physics, then the inevitable conclusion is that a

400 or 800 ton UFO has an inertial mass of 1 kg . This is not THEORY. It is FACT.

In our world, as we move objects about, the force of inertia can be felt, but it is small. When we accelerate space craft, the forces can be huge. As mentioned above, at one point in its ascent, the Shuttle goes through an acceleration of 9g. What this means is that the force applied to the Shuttle to overcome inertia is nine times its weight, millions of pounds.

The reason that UFOs can flit about, making such incredible moves, is that they are not hampered by inertia. Since the inertia they experience is about one millionth of what our space craft feel, they need practically no energy to do what they do.

In brief,

The Second Law of Thermodynamic implies:
silent motion = no energy use
no energy use = no inertia
no inertia = inertial mass (essentially) zero.

The Law of Conservation of Momentum is violated by sharp angle turns, unless the momentum before and after the turn are both zero. UFOs CANNOT make smooth turns (as we shall see in the chapter on "UFO Propulsion"). They can only do sharp angle or "zig-zag" turns. And "zig-zag" turns mean inertial mass is near zero.

(For those of you who have heard the Dr. Einstein insisted the gravitational and inertial mass to be the same for General Relativity, you must remember that Einstein considered a universe of positive gravity only. As far as I know he never dealt with negative mass. And there lies the difference. We will discuss this at length in the chapter on "A Dissertation on Mass".)

1) Imbrogno, Phililp I. ***Contact of the 5th Kind***, Llewellyn, St. Paul, 1997, p.39

5 Unidentified FLOATING Objects

There have been estimates that there are three million sunken vessels under the waters of the world. It is a sad fact, even without calamities, natural or man made, that if you simply ignore a hull, it will sink, even if its just because of rain water.

Boston Whaler boats have unsinkable hulls. If you cut a Boston Whaler hull in half, each half will float. Now suppose all boats and ships had hulls which were built like Boston Whalers (most small boats now are). Then there would not be 3 million sunken hulls under the sea. If a vessel became disabled, it would merely drift till it washed up on shore, or be towed to port, no matter what condition its hull was in.

There have been at least a million UFO visitations on Earth during the last century alone. There are over 209,000 eye witness reports in the ***UFOCAT*** database alone, and we know that not all UFO sightings are reported. The earth is not pockmarked by UFO crash sites. There is case in Rosewell New Mexico in 1954 that has been debated forever. The "autopsy" film purported to be from the event has been shown to be a fake (1).

There is a new book that discusses UFO crashes by Kevin Randle, ***Crash: When UFOs Fall From The Sky*** (1). The problem is that there is practically no evidence of debris. There is apparently no photograph of a crashed UFO , only tiny bits of matter. Actually the term "crash" may be inappropriate, as it implies the UFO plummeted to Earth by gravity. Several of the cases discussed are actually of UFOs entering the water and eventually taking off again. The ***UFOCAT*** has a dozen reports of UFOs doing just that. Also the author himself says that a large portion of the cases of crashes he discusses are hoaxes. We could not find anything in the book that seriously contradicts what we propose.

UFOs are machines. Machines break down. If they need to be kept aloft by some active mechanism, then that mechanism is likely to fail given a million opportunities. So the absence of UFO crash debris probably has another explanation. One simple explanation is that UFOs naturally float in gravity, and if they are disabled they can somehow limp home without crashing. Being buoyant, they do have to do anything actively to counteract gravity.

We have the example of lighter than air craft, namely balloons and blimps. If a blimps' motor fails, it will just drift without crashing, just as balloons drift. But the structures of blimps and balloons is frail. They can not withstand the forces generated by rapid movement in air. UFOs, on the other hand, have been clocked by radar at well over 1,000 miles per hour, and we do not know what the limits of their speed are. Because they do reflect radar we know that they are materially substantive. And people who have seen them close up report that they appear to be made of metal. Therefore all indications are that UFOs are massive. They are not buoyant because they are lighter than air. Therefore something else must be involved.

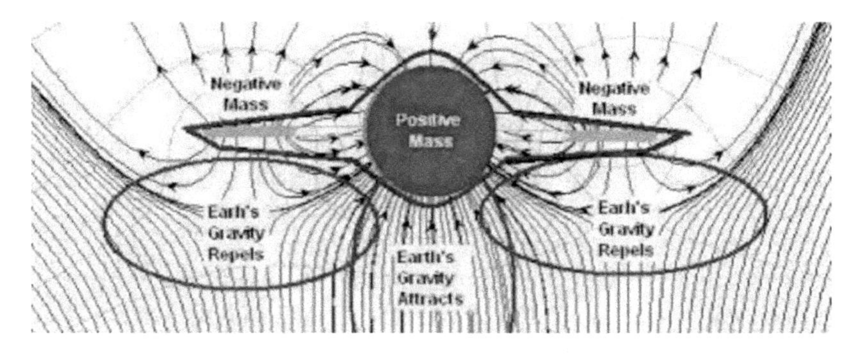

And that something else is negative mass. Were the question merely that of buoyancy in gravity, it would be more difficult to conclude that the buoyancy was due to negative mass, but that is not the case. We also have already seen their ability negate inertia and will discuss the ability propel themselves with anti-gravity propulsion shortly, both of which also can be achieved with negative mass. Further more, a geometry exists where this negative mass is concentrated in a single movable part which can achieve all of the major UFO activities. It is this fact which re-enforces the idea that UFOs float because of negative mass.

UFOs have equal amounts of positive and negative mass in their craft. The pull of Earth's gravity on these masses is the same as the push upward, resulting in a net zero effect on the craft.

It is the positive gravitational charge of ordinary objects which interacts with Earth's gravity to produce weight. But if you have equal amounts of positive and negative mass, the forces of attraction and repulsion are equal and the craft "floats" in gravity.

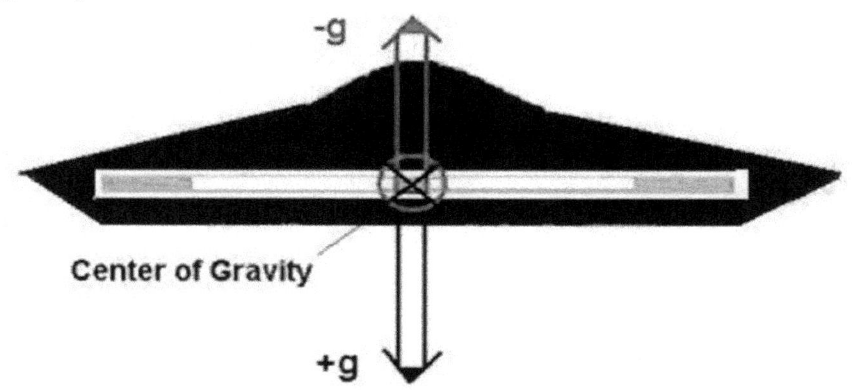

During the Hudson Valley and Belgian UFO waves there were literally thousand of eyewitness reports that the Big Black Triangle UFOs floated slowly, even at walking speed, and in utter silence over the landscape at low altitudes. People standing directly below sometimes heard a faint hum they said sounded like a vacuum cleaner (2). Authorities bent over backwards to explain that the sightings were of ultra light aircraft. They even sent up such craft to fly in formation to simulate UFOs (3). Eyewitness, of course, did not buy these explanations.

Another problem is the silence. We know that balloons and blimps can float silently, but they then can't take off at hundreds of miles an hour and disappear, like UFOs regularly do. The simple fact is that by the **Second Law of Thermodynamics**, if you use energy to stay aloft, the energy must appear somewhere as noise or turbulence. To have the Venturi effect on wings lift a plane up, engines must be used to move the plane through the air, and the engines make noise, create prop wash, and produce turbulence. The utter silence of UFOs means that they are not using energy to stay aloft. (The energy indicated by the faint hum is not enough.) That means that they must be buoyant, which does not require energy. And being apparently made of metal and heavy, the only way they can do that is to neutralize the pull of gravity with equal amounts of gravitationally positive and negative mass.

1) Randle, Kevin D., Crash When UFOs Fall From The Sky, Career Press, Franklin Lakes, N.J., 2010
2) Hynek, J. Allen; Imbrogno, Philip; Pratt, Bob; Night Siege, Llewellyn, St. Paul, MN, 1998, page 32.
3) Ibid., page 56.

6 Gravitational Engineering

Summary: **We can't do Gravitational Engineering. UFOs can .**

UFOs are able to create huge gravitational fields, rather anti-gravitanional fields, that can outlift the Earth. They can do that because they have gravitational engines consisting of one moving part. It is a ring of gravitationally negative mass. The huge anti-gravity fields are created by spinning the negative mass furiously, at almost the speed of light. The chief reason they do that is to become inertia free, and to propel their craft. But before we get into that, let us first go back to the beginning, and see why we can't do what they do.

The reasons are simple. There are basically three levels of magnitude involved in gravity.

The first is the size of ordinary objects we deal with, say from one to thousands of kilograms. Each of these objects produces a very small gravitation field which we will call miniscule.

The next level of magnitude is the astronomic. As an example take the Earth. The Earth has a mass, **M, of 6.6 x 10^{24} kilograms**. For purposes of gravity, we can consider this mass to be concentrated at the center of the Earth, so on the surface the distance to center is **6,000 kilometers**. It creates an acceleration on the surface of

$$\mathbf{g = GM/r^2}$$
$$\mathbf{= 9.8\ m/s^2}$$

where **G** is the gravitational constant.

The miniscule gravitational field of our object interacts with the astronomic gravitational field of the Earth, producing something measurable, namely the weight of the object. Let's say the object is 1 kilogram, then the weight is a force, and the force is given by

$$\mathbf{F = mg = (1\ kg) \times 9.8\ m/s^2}$$

$= 10$ **newtons.**

Now suppose we want to find out what the gravitational force would be between two of our objects. For **two 1 kg objects 1 meter apart**, the force would be:

$$\mathbf{F = Gm_1m_2/r^2}$$
$$\mathbf{= (6.7 \times 10^{-11})(1\ kg)(1\ kg)/\ (1\ meter)^2}$$
$$\mathbf{= 6.7 \times 10^{-11}\ newtons.}$$

This force is essentially immeasurably small.

The third gravitational scale is the cosmic. Estimate of the total mass of the universe centers around $\mathbf{10^{55}}$ **kilograms**. Integrating over this entire mass produces the universal background scalar potential $\mathbf{\Phi}$.

The gravitational field can be decomposed into two parts, the Newtonian Gravitation, $\nabla\varphi$, and an inertial term $\partial \mathbf{A}/\partial t$. $\mathbf{\Phi}$ and **A** are not fields. They are potentials. Potentials are mathematical precursor structures from which gravitational fields can be derived.

As we stand on the Earth, we are immersed in the Earth's gravitational field, which of course is not visible. If we hold out an apple in front of us, the **minuscule gravitational field** interacts with the **astronomical gravitational field** of the Earth. The interaction between the two is one of attraction. The apple pulls against our hand. If we let the apple go, the attraction between the apple and the Earth will make the apple fall. Therefore the Earth's gravitational field has palpable effects.

The potentials φ and **A** do not have palpable effects. It is only the gradient, or space derivative of φ, and the time derivative of **A**, which are fields.

The gravitational field can therefore be decomposed into derivatives of the precursor potentials

$$\mathbf{E} = -\nabla\varphi - \partial \mathbf{A}/\partial t.$$

A scalar potential must have a curvature or gradient, ∇, to produce an actual gravitational field **E** and thus a gravitational force . The gravitational potential in space, **Φ**, is flat. It has no gradient. Therefore objects in space are weightless.

The only experience we have of the vector potential **A** is in context of inertia. As is discussed in the **ARTICLE**, both Sciama and Martin et. al. considered a universe receding with velocity **v**. Then integrated over the mass of the entire universe to arrive at the vector potential **A**

$$\mathbf{A} = \int \rho \mathbf{v}/r \, dV$$

Since the entire universe was moving with constant velocity -**v**, **v** came out of the integral, leaving

$$\mathbf{A} = \mathbf{v}\int \rho/r \, dV$$
$$= \mathbf{v}\Phi/c^2$$

They then switch coordinates, and have the universe stationary, and the particle moving with velocity **v**. Therefore the only measurable vector potential that we know, comes from integrating of the entire mass of the universe. Astronomical amounts of mass, or miniscule do not produce vector potentials **A** that we can detect.

We then can interpret this that the miniscule field of an ordinary object moving against the scalar potential **Φ** produces the vector potential **A.**

This **A** has no effect. But if the object is ACCELERATED against **Φ**, a **$\partial A(-)/\partial t$** is produced, and this is gravitational field, **E**. The accelerated body, **m**, experiences a reactive fore, and this reactive force is the force of inertia. This is a measurable force which we see at work every time we drop an object. Inertia slows its fall; without inertia the object would fall at speed of light. It exists only because of **Φ**, the summation of the potential of all the matter in the universe.

So the measurable gravitational forces we deal with ordinarily are the weight of objects which is

$$\mathbf{F} = \mathbf{mG}\nabla\varphi$$

and the inertial force on accelerated objects

$$\mathbf{F} = - \mathbf{mG} \, \partial \mathbf{A}(-)/\partial t$$
$$= - \mathbf{mGa}\Phi/c^2$$

The reason for the terminology is interesting. Einstein's full gravitational equations are notoriously difficult to calculate with. But it turns out that if you are not interested very strong gravitational fields like around black holes, or velocities near the speed of light, then what are known as the "weakfield approximations" to the full gravitation fields are sufficient.

These weakfield approximation equations are IDENTICAL to Maxwell's Equations for electromagnetism. That is fortunate, because Maxwell's Equations are the most studied equations in science. Both physicists and electrical engineers have developed solutions to every conceivable configuration of problems over the last century. If the approximation is applicable, all you need to do is to look up the corresponding solution to electromagnetic problem and substitute and you have the gravity solution immediately.

Therefore it is now customary when doing gravitational calculation just to use even the electromagnetic symbols for the variables. So the gravitational field is $\mathbf{E}$, the gravitational scalar potential is $\boldsymbol{\varphi}$, and the gravitational vector potential is $\mathbf{A}$, with not even subscripts to denote that one is dealing with gravity.

The Newtonian gravitational force is produced by a gravitational charge which is essentially the mass. There is nothing we can do about the mass, except accumulate it. There is no hope of amplifying the Newtonian field through engineering.

Therefore to do any kind of gravity engineering, would have to be on the inertial term or the vector potential $\mathbf{A}$.

How do we amplify with electromagnetism? Electric charge is not fixedly attached to mass. Electrons in the conduction band of a metal move about so freely, that in physics they can essentially be treated like a free gas. Apply an electric field, and the electrons move (drift) and in large numbers we call "currents". The mass of electrons compared to other components of matter is so small as to be negligible. The currents thus generated can be sent through coils of wire to produce magnetic fields, and otherwise manipulated and their effect multiplied.

Matter also has a charge, a gravitational charge. Except the gravitational charge of matter is about $\mathbf{10^{-42}}$ times smaller than the electric. Gravitational charge does not decouple from matter. To move the gravitational charge we must move the matter itself. Gravitational charge cannot be sent through a wire.

Can one do anything to magnify gravity? The answer is yes -- but.

If we take an electrical charged disk and spin it, it produces a magnetic dipole. And the dipole in turn produces the magnetic vector potential **A**. **A** is defined by,

$$\mathbf{dA_e = J_e\, dV}$$

where $\mathbf{dA_e}$ is an element of the vector potential, $\mathbf{J_e}$ the electric

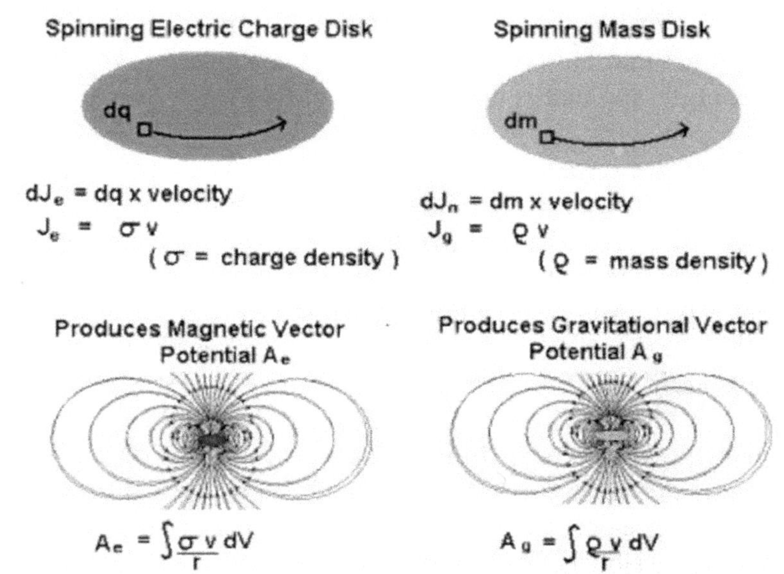

Similarly, if we take a spinning disk and consider its mass, then a portion of the disk, **dm**, then moves around. This also creates

a current, and **dV** an element of volume. Consider a portion of charge, **dq**, on a charged disk. As the charge on the disk moves around, the charge creates an element current $\mathbf{dJ_e}$. As the element **dm** moves around it creates current element, but of mass of, $\mathbf{dJ_g}$, is $\boldsymbol{\rho}$ the mass density, and $\mathbf{J_g}$ the mass current.

The spinning disk then produces a gravitational vector potential $\mathbf{A_g}$

$$\mathbf{A_g} = \int \boldsymbol{\rho} \mathbf{v/rdV}$$
$$= \int \mathbf{J_g/rdV}.$$

While we can't separate the gravitational charge from the mass, by moving the mass we can create a mass current. This mass current then will induce a GRAVITATIONAL vector potential $\mathbf{A_g}$. This vector potential is very small.

Can we amplify this gravitational vector potential $\mathbf{A_g}$? Well, we have several things at our disposal. We cannot increase the mass density $\boldsymbol{\rho}$. In nuclear physics we can make transuramic elements which are somewhat heavier than uranium, but they are unstable. So we cannot increase the mass density. We can therefore try to increase the total mass we spin, or the speed of the spin. The amount of mass we use will depend on how much we can get into our craft, if we think of using the gravity in a space craft, and that is limited.

Therefore we have to increase the velocity of the spin. But our spin is unfortunately limited by physics and properties of matter. Elements of a rotating disk are constantly changing direction as they rotate. A change of direction is an acceleration. And by **Newton's Second Law** a force is need to for this acceleration. The force in case of the spinning disk is provided by the material strength of the matter of the disk.

A centrifugal force tries to keep the element of the disk going straight. As the disk rotation speed increases, the centrifugal force puts greater and greater stress on material of the disk, until it invariably disintegrates.

Take the case of Uranium separation centrifuges. They are

spun in the neighborhood of 60,000 revolutions per minute. They are only about 6 inches in diameter. But at higher speeds the centrifuges disintegrate.

In order to magnify our $\mathbf{A_g}$ by a factor of 10^{10} or more, we would have to spin negative mass, several meters in diameter, at over 1,000,000 revolutions per second so we are unable to amplify gravitational fields to a magnitude to have any practical use. Any material we try to use would just disintegrate.

Then how can gravity be amplified?

If the mass we were spinning had NEGATIVE gravitational charge, something else would happen. As the negative charge spun, it would create a negative vector potential $\mathbf{A_g(-)}$. This negative vector potential repels the positive vector potential $\mathbf{A_g(+)}$ that is responsible for inertia. It is the motion of an object in space immersed in the universal background scalar potential $\mathbf{\Phi}$ that creates the positive vector potential $\mathbf{A(+)}$ that is responsible for inertia and centrifugal force.

The $\mathbf{A_g(-)}$ generated by the spinning negative mass disk would create a bubble in which there were no inertia. The negative disk would no longer be subject to inertia. The **centrifugal force** that ordinarily tears a fast rotating disk apart is an inertial force needed to constantly change the motion of an element of the disk. If the disk had no inertia, no force would be needed to constantly change direction of that element of the disk. There would not be a **centrifugal force**. Therefore the disk could be spun at any speed, up to where the rim velocity approaches the speed of light.

Mach's Principle says that the "distant stars" (or $\mathbf{\Phi}$) create inertia. The corollary to Mach's Principle is: if the effect of the "distant stars" (or $\mathbf{\Phi}$) is excluded, there is no inertia.

So the question of doing gravity engineering comes down to the fact whether or not one has negative mass. We, unfortunately, don't. It appears UFOs do. Much of the evidence is indirect, but cumulative. However there is some very direct evidence. If you do the vector analysis of how, for example, a UFO can wind up lifting a car off the road, the vector analysis points directly to the fact that negative mass is involved. We will discuss this shortly.

7 How Levitation Points to Negative Mass in a UFO

Most of the three hundred reports of levitation also mention loss of car steering control. What that means that is the levitation takes some of the load off the cars tires. They loose traction. The car will not follow where the front wheels point. Lifting the car of the road is just an extreme example, where the levitating force exceeds Earth's gravity.

Therefore that UFOs are able to generate palpable gravitational fields. ANY "levitation" proves this. We can't make gravity. UFOs can. The lifting of a car is merely a spectacular example.

What the lifting of a car means that the fields UFOs are able generate are GREATER than Earth's gravity. When the car was lifted off the ground, the UFOs gravity field and the Earth's gravity field pulled in opposite directions and the UFO field won. That means that a UFO, about 30 feet in size could outlift the Earth which has a mass of 10^{24} kilograms.

The first thing to note is that a stationary UFO does not exert any gravitational force. The main purpose of its mass ring gravitational engine is to produce a huge negative vector potential field to block any positive inertia generating fields and to propel the craft. The vector potential field it produces is constant. The rotation speed of the ring probably never varies.

Gravitational forces are produced by CHANGES in the vector potential field **A**.

$$\mathbf{F} = -\,\mathbf{mGE} = -\,\mathbf{m\,G}\partial\mathbf{A}(-)/\partial\mathbf{t}$$

It is therefore the MOTION of the UFO that produces a gravitational field by changes in the huge vector potential **A(-).**

The gravitational vector field **A(-)** from the negative mass ring is in the form of a torus, with **A(-)** vector parallel to great circles on the torus. We will take a representative vector **A**

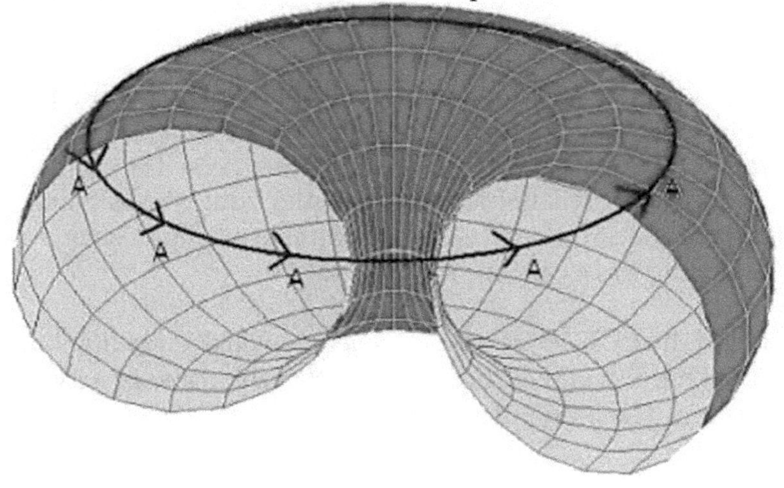

horizontal just in front of us in the middle.

If the UFO were spinning a positive mass, which it is not, then the situation would be as follows. The vector potential at the car would be increasing as the UFO approached. The change in the vector potential is made up of the increase due to the approach, times the velocity **v**.

$$\frac{\partial \mathbf{A}}{\partial \mathbf{t}} = \frac{\partial \mathbf{A}(+)}{\partial \mathbf{r}} \mathbf{v}$$

It is this change in the vector potential which creates the gravitational force

$$\mathbf{F} = -\ \mathbf{m}\ \mathbf{GE} = -\ \mathbf{m}\ \mathbf{G}\frac{\partial \mathbf{A}}{\partial \mathbf{t}}(+).$$

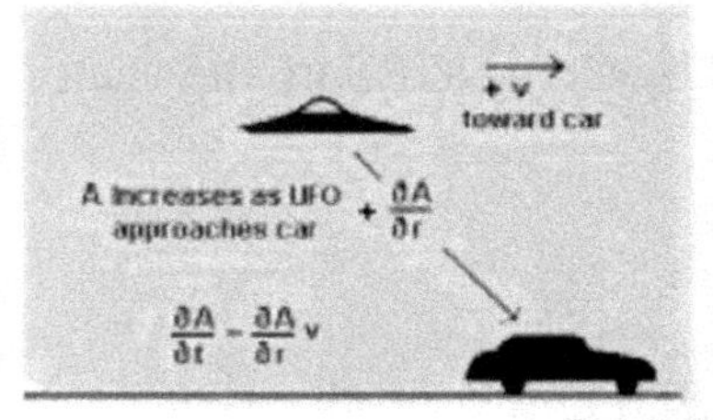

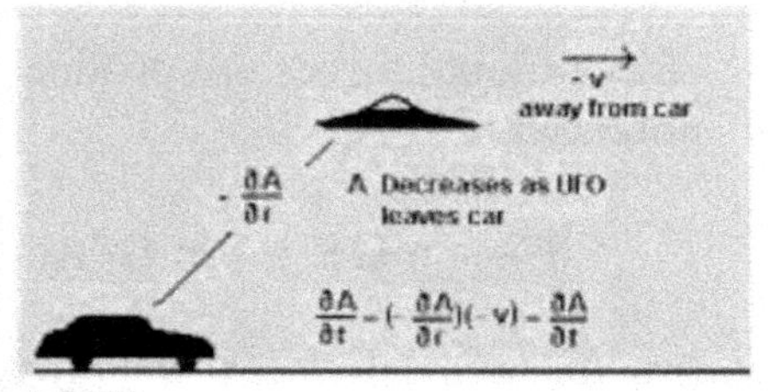

Gravitational Field
$\frac{\partial A}{\partial t}$ the same coming or going

As the UFO approaches, **A(+)** increases, so **∂A(+)/∂r** is positive, and so is **v**. As the UFO departs, **A(+)** is now decreasing, so **∂A(+)/∂r** is decreasing, and the vector **v** is departing, therefore negative with respect to the car. Since both **∂A(+)/∂r** and **v** are negative, the product of two negative is a positive. Therefore with respect to the car, **∂A(+)/∂t** the same sign whether the UFO is approaching or departing.

Note that with the positive gravitational ring we have assumed for argument, the **∂A(+)/∂r** is TOWARD the car. Hence the resulting F is toward the car. Any gravitational forces created would be PUSHING THE CAR TO THE GROUND.

Therefore the ring in the UFO has to have NEGATIVE GRAVITATIONAL MASS, then vector potential **A(-)** it produces is OPPOSITE in sign to the **A(+)** of a positive mass. Therefore the directions of the vectors is reversed. **∂A(+)/∂t** now points AWAY from the car. The gravitational force it creates now LIFTS the car.

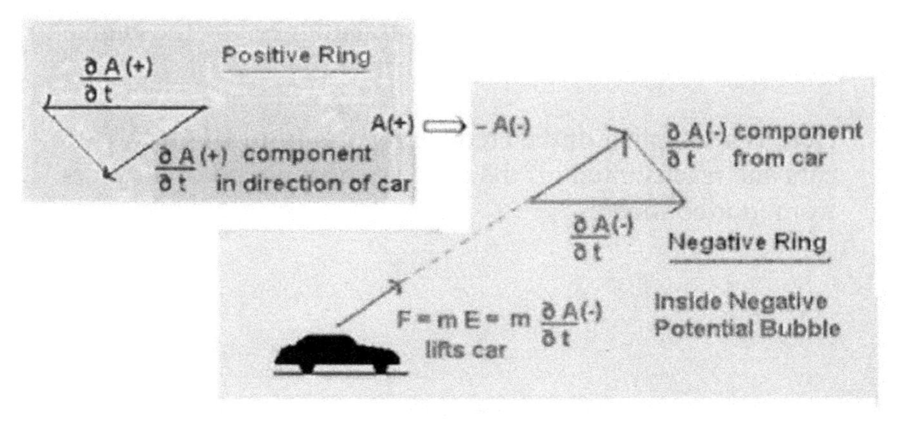

Note that the lifting force is usually not straight up. The lifting force is a component of the **∂A(-)/∂t,** of the craft, which is passing, probably horizontally. We can see from the torus diagram that the lifting forces on the car can come at various angles as the UFO passes, we would expect that they would effect the lift in different points. And this is indeed is what eyewitness report. Sometimes only back or front of the car is lifted, sometimes the car is lifted on two wheels. But sometimes the whole car is lifted off the ground and dropped with a thump, as the UFO leaves, occasionally bursting a tire.

The lifting force is only momentary. The value of **A(r)** falls of as the UFO moves away. The lifting force diminishes and the car drops to the ground.

So a UFO lifting a car off the road is an unintended side effect of the large gravitational vector potential it produces to nullify inertia and propel the craft. It is like the blast of air from rapidly moving truck, that blows you around on the highway unintentionally.

But we see from the vector analysis of the forces involved, that a LIFTING force implies that the spinning ring in the UFO is gravitationally NEGATIVE mass.

What is even more interesting, as we see in the **ARTICLE**, that the strength of the negative vector potential **A(-)** needed to lift the car is IDENTICAL to the strength need by the engine to overcome inertia.

Here is a sampling of the credible reports from **UFOCAT** of cars and trucks lifted off the road during a period of 40 years from four continents.

PRN	YEAR	STATE	NOTES
034033	1960	Ven	Disk color of polished blue steel swooped down very close to hood of truck and lifted it 1 m above road.
109603	1965	Bel	Yellow BOL very close to windshield car wheel not responding, lifted some cm off ground for 3-4 seconds, 50 m on road.
050879	1969	IL	Object shaped like rounded triangle came very close to car, lifted car 10' off road. Motor and lights continued running, but steering failed until UFO left.
092874	1971	AB	NL paced the car, car lifted 2' off road, set down with bump.
083928	1973	NC	Domed disc chased and lifted car off road. 2 feel weightlessness, no control of vehicle.
109415	1978	MO	Silent NL lifts rear end of car, then drops it. Independent observer.
109437	1978	ARG	Brilliant yellow UFO approached car, lifted car off road.
156554	1978	ARG	Father and son, both mechanics, driving in 1930 Chevy pickup, car lifted from road, engine stopped, car stereo stopped.
192536	1979	ARG	2 men in final leg of 39-day car race in Citroen. Car lifted some 6-8' above road.
119737	1980	CO	EM effects: lights dimmed, radio out, car lifted.
119573	1981	FL	2 lights - large silver UFO, lifts car off road.
118750	1981	TX	Domed disk hovered over truck, forward motion impeded, truck lifted off road.
120458	1983	PA	Silver disc passed low over car, wheels lifted off road.
118115	1984	PA	Woman in car sees 24' disc overhead, car lifted on 1 side for 3 seconds, then dropped!
114024	1986	WAU	Oval object buzzed car, lifted car off highway. Damaged tires and car badly.
120871	1992	CA	3 events, car lifted off ground, missing time.
191566	2001	NSW	Near collision with vehicle disc, bright white, lifted car whilst waiting for traffic lights in Sidney, Australia.

8 How to Kill Inertia

IMMERSION

An aircraft flies through the air. A boat travels on water. The motion of the aircraft and the boat are retarded by the resistance of the fluids to their motion. Power must be applied to overcome the air and water resistance. Both air and water resistance are essentially proportional to the velocity of the motion. The faster you want to go, the more power you have to apply.

A space craft is immersed in the gravitational background scalar potential **Φ** which permeates the universe. This potential does not offer resistance to motion. Once launched, we know an interplanetary rocket can coast to its target for years. But the background scalar potential offers resistance to ACCELARTION. If you want to speed up the craft, you have supply force. That is **Newtons's Second Law, F = ma**. You have to overcome inertia. Inertia is caused by the scalar background potential which produces inertia through the vector potential **A(+).**

The UFO is made of equal amounts of positive and negative mass. The time derivative of the vector potential **A(+), ∂A(+)/∂t,** acting on the positive mass produces a gravitational field which is the reactive field of inertia. It is a direction opposite to the acceleration. But the negative mass produces a "perverse" inertia in the direction OF the acceleration. Since the positive and negative masses are equal, the inertial field of the positive mass and "perverse" inertial field of the negative mass are opposite and equal, they actually cancel each other out.

But there is a problem. They produce immense stresses on the UFO. For the 35g acceleration we have been considering, even for a small UFO with 40 tons of each mass, the pull in opposite directions of the inertial forces would be 5 million pounds.

The universal scalar potential **Φ** permeates all matter. It produces the vector potential **A(+)** inside matter. Upon acceleration the **∂A(+)/∂t**

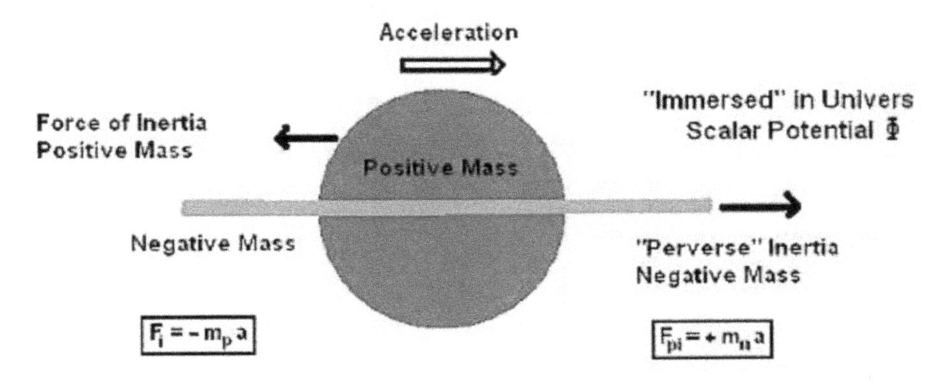

generates a gravitational field which interacts with the miniscule field of the object. The inertial force is exerted at the point in the object. The strength of the force is proportional to the mass of the object. So for example an 150 lb occupant of the UFO, under a 35g acceleration, would feel an inertial force of 5,000 lbs. You do not want to subject the occupants of the UFO to such, possibly lethal, forces. Therefore you need to eliminate this inertial force.

If the plane was not immersed in air, and the boat in water, there would be no resistance. So to avoid inertia, one must eliminate the immersion in the in the vector potential **A(+)** that is generated by universal scalar potential **Φ**. That is what UFOs do. With gravitationally negative mass, they create a bubble in space that travels with the UFO. This bubble excludes the vector potential **A(+)** and **∂A(+)/∂t** which cause the force of inertia. Therefore the UFO does not feel the resistive force of inertia.

To explain in more detail: the gravitational potential **φ** of all matter in the universe adds up to a grand background scalar potential **Φ. Φ** pervades the universe. If any object moves against this **Φ** a vector potential **A** is created.

$$\mathbf{A} = \mathbf{v}\,\Phi / c^2$$

Under ordinary circumstances this has no effect. But if the object is ACCELERATED, then there is generated a time derivative of the vector potential **∂A/∂t**, and this time derivative creates a gravitational field **E** which is the reactive field of inertia.

$$\mathbf{E} = -\frac{\partial \mathbf{A}}{\partial \mathbf{t}}$$

If a body moves against the scalar potential Φ ⟶ A(+) is produced. This A(+) has no effect.

$$A(+) = v\frac{\Phi}{c^2}$$

If body is ACCELERATED against Φ there is a $\frac{dA(+)}{dt}$.

$$\frac{dA(+)}{dt} = a\frac{\Phi}{c^2}$$

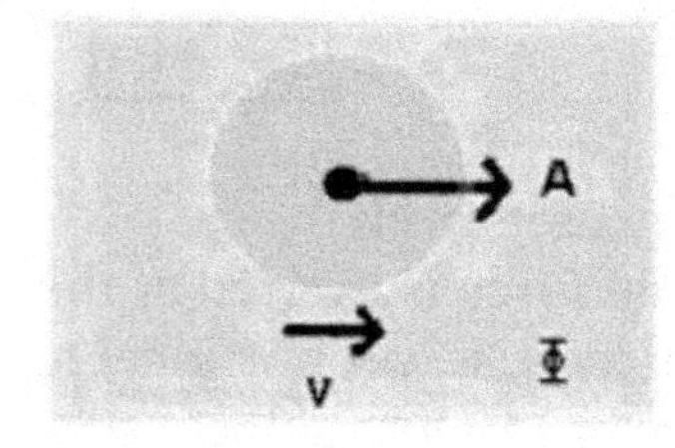

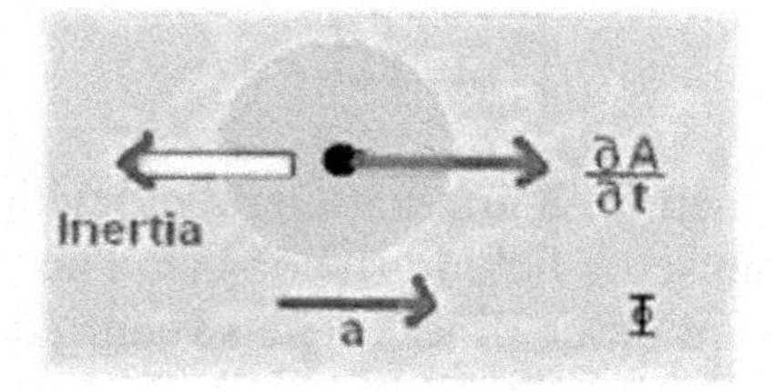

$m\frac{dA(+)}{dt}$ is a force. It is a REACTIVE force.

It is INERTIA.

This reactive field **E** then generates a force

$$\mathbf{F_i = - m\,GE \;=\; -m\,G\frac{\partial A}{\partial t}}$$

which is the force of inertia. This is in **Newton's Second Law: F = ma.** Since the object is immersed in the background scalar potential **Φ**, which is everywhere, it can not escape being subject to inertia. The only way one can escape inertia is to dig a hole in the vector potential **A**, and in effect hide from it. One can not hide from **Φ**, but it is possible to hide from the vector potential **A(+)** that it produces.

And this is what UFOs do. The **A(-)** vector creates a bubble in which there is no inertia. This bubble travels with the UFO and the UFO hides in it.

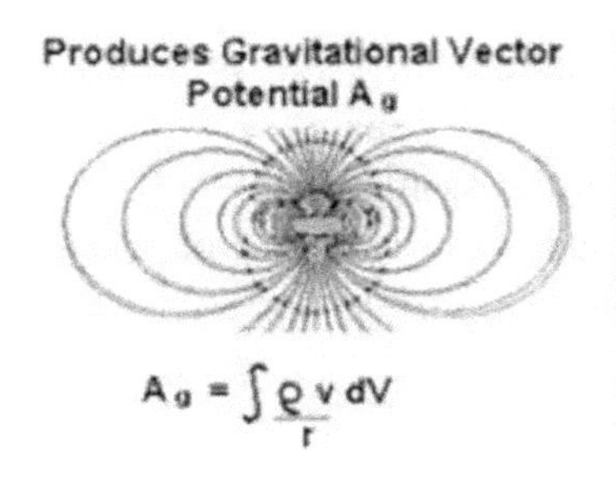

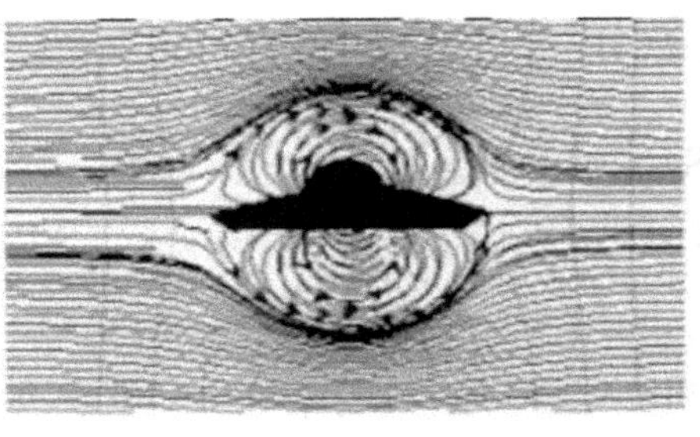

We do not know of, or have ever seen, gravitationally negative matter. All matter in the universe, as far as we can tell, is gravitationally positively charged. But apparently UFOs have it.

If you integrate over all the matter in the universe, you get the following relationship: An object moving in space, that is, in the background potential **Φ**. will generate a vector potential

A = velocity/G

where the G is the gravitational constant. This result comes from Sciama (1) and is derived in the **ARTICLE** "Fundamental Equation of a Flying Saucer."

Note that **A** is velocity dependent. Suppose we pick a simple number: velocity = 1 km/sec or 1000 m/sec. This velocity corresponds to 2,200 miles per hour. From the Belgian radar data, such speeds MAY have been recorded for UFOs. Suppose the UFO would like to be free of inertia while traveling up to such speeds. The vector potential **A** at this speed would be

$$\mathbf{A = v/G}$$
$$\mathbf{= 1000\ m/sec\ /\ 6.6 \times 10^{-11}}$$
$$\mathbf{= 1.5 \times 10^{13}\ kg\ sec/m^2}$$

Any object moving at this speed would be subject to such a vector potential.

A gravitational negative mass would have negative fields around it, but they would be puny, as is the gravitational field of any finite object. In essence, the negative field would confined to the mass itself. It could not create a hole in the huge positive scalar potential.

But the definition of A contains the velocity **v:**

$\mathbf{A_g} = \int \rho \underline{\mathbf{v}}/\mathbf{r}d\mathbf{V}$ where ρ is the mass density.

Suppose we now spin the mass. Then the spinning gravitationally negatively charged mass would start to create a negative gravitational vector field, that of a gravitational magnetic dipole. As we spun the mass faster and faster, the dipole and the resulting vector potential would get larger.

Due to the spin, the negative vector potential **A(-)** will increase. Positive and negative gravitational fields repel each other. As we show in the **ARTICLE**, these are mutually exclusive. Therefore **A(-)** is SUBTRACTED from **A(+).** In other words the **A(-)** from the spinning mass digs a hole in the **A(+)** arising from the universal

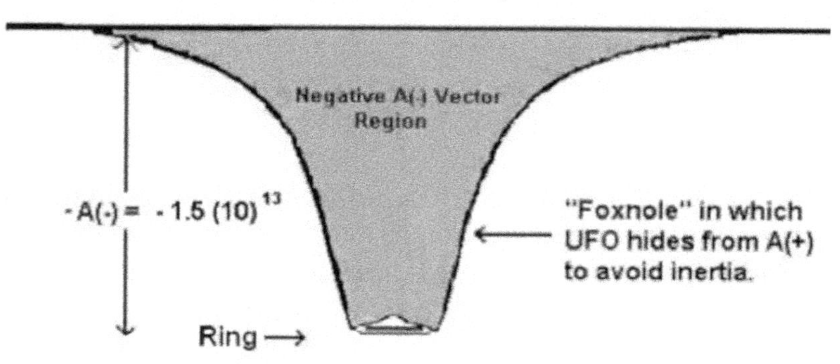

background potential **Φ**. It turns out, as we will see later, that the negative mass has to be in the form of a ring (like a Frisbee) for enough vector potential to be created.

If the UFO is not moving, there is no **A(+)** generated by motion against the scalar background scalar potential. The spinning negative ring just creates the negative **A(-)** vector potential. The UFO sits in the "foxhole" well created by the disk.

If the UFO is moving, then a positive vector potential is created, and we have to add the positive **A(+)** to the negative **A(-).** If it were to dig a hole DEEPER than the **A(+)** generated by the moving body, then at all times, whether it was moving or not, it would always be in the negative potential **A(-),** it would never be SUBJECTED to the positive vector potential **A(+),** which causes inertia (upon acceleration).

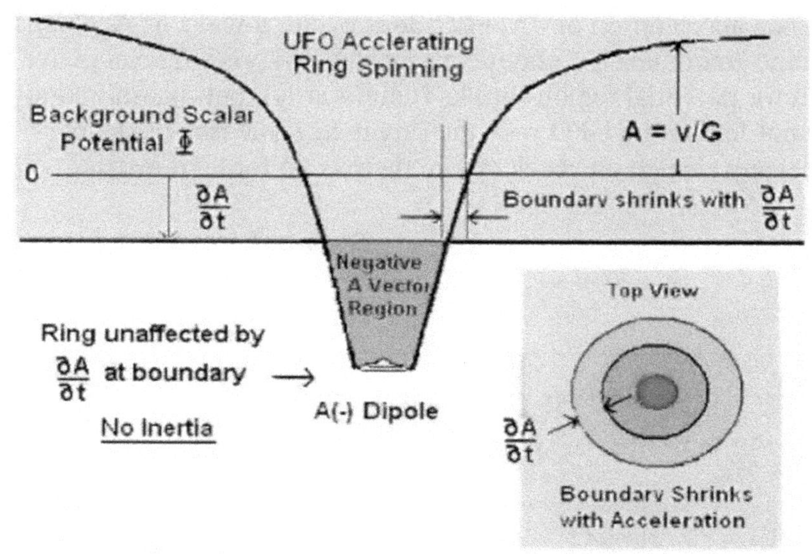

If the UFO now accelerates, the **A(+)** that it generates would increase. There would be, during the acceleration a **∂A(+)/∂**t. The absolute level of **A(+)** would shift. But if the UFO were sitting inside a foxhole dug by the spinning negative ring which is DEEPER, then it would not feel any changes due to accelerations.

All the accelerations would do is shift the **A(+)** = **v/G** up or down. This would result in shrinking or expansion of the boundary between the **A(+)** and the **A(-)**. But the **A(-)** field at the ring would be constant, determined by the spinning of the ring, and nothing else. Since the **A(-)** is constant, there is no **∂A/∂t,** - no inertial force, no inertia.

Another way of looking at it is to invert the diagram of the vector potential. Here the **A(-)** potential "foxhole" dug by the spinning ring would be seen as an island mountain rising from the sea of the positive vector potential **A(+).** The force of inertia only arises when there is an acceleration, or a change in the vector potential **A(+),** that is **∂A(+)/∂t.** The spinning ring creates a constant **A(-).** It does not directly neutralize **∂A(+)/∂t.** What it does is put the ring out of reach.

When acceleration or **∂A(+)/∂t** does occur, a wave of **A(+)** laps at the "mountain", without ever reaching the ring. The negative vector potential region shrinks (temporarily), but this shrinking is not felt by the UFO with the ring in it. If the ring does not feel any change on acceleration, there is no force of inertia.

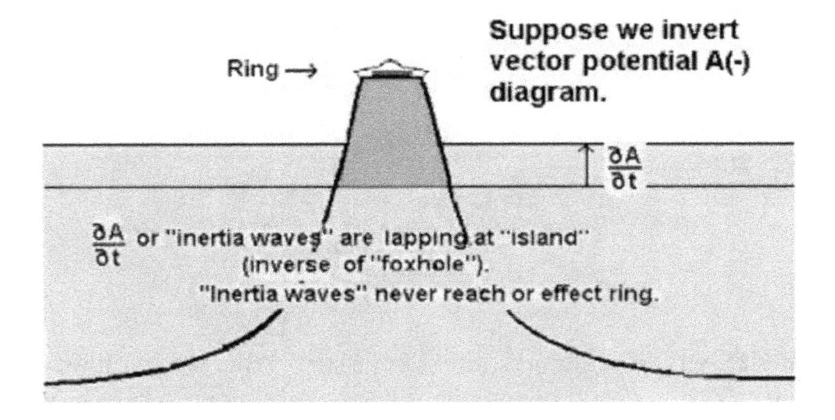

Therefore, the only question is, can the UFO with its negative mass spinning ring dig a foxhole deep enough to hide in. Can it generate a negative vector potential grater than **A(-) = 1.5 x $\mathbf{10^{13}}$ kg sec/$\mathbf{m^2}$**. We will deal with that in the next chapter on the UFO's gravitational engine.

9 The UFO Anti-Gravity Engine

There are apparently many types and shapes of UFOs. But the vast majority appear to be of the saucer type. Indeed, there are hundreds of photographs like the sketches below.

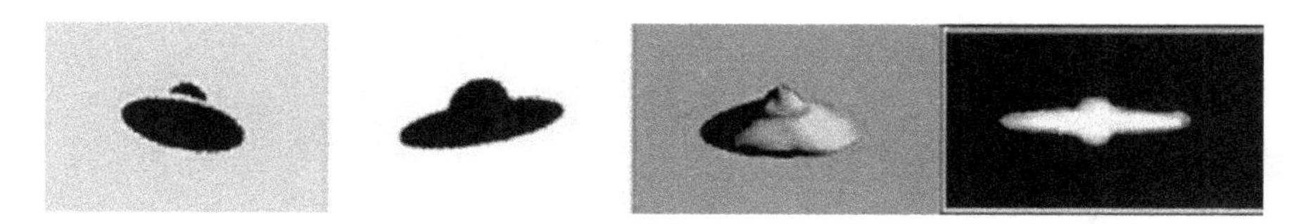

Closely related to these are Big Black Triangles. As we shall see, these are most likely "multi-engine" versions of the saucers. It is about these two types that we have the most detailed information. It is for these two types that we will develop our explanation. I suspect that other versions share the same structure, but that is beyond the scope of this book.

All UFOs have anti-gravity engines which have two purposes:

1. Create a negative vector potential field to block the positive vector potential from the universal background potential that creates inertia.
2. Provide that gravitational force that is the motive force which propels the UFO in space.

To arrive at nature and strength of the UFOs gravitational engine we have some criteria we have to meet. It turns out that size is the crucial element. It is much more difficult to design a small engine that will do the job. Why that is, we will see shortly. From the welter of eyewitness reports, it appears that a thirty foot diameter saucer is the smallest that is seen. The engine in a such a saucer must be strong enough to block inertia. And that is a very tight constraint. It will in essence determine just about every detail of the engine.

It is possible that not all UFOs actually meet all the requirements that we have set. It is possible that only larger saucers can be inertia free at 2,000 mph for which we MAY have data from Belgian radar. Maybe smaller saucers can't. Just to be sure we will require our UFO, a small saucer, to meet all the requirement, and see we will try to see if can construct one.

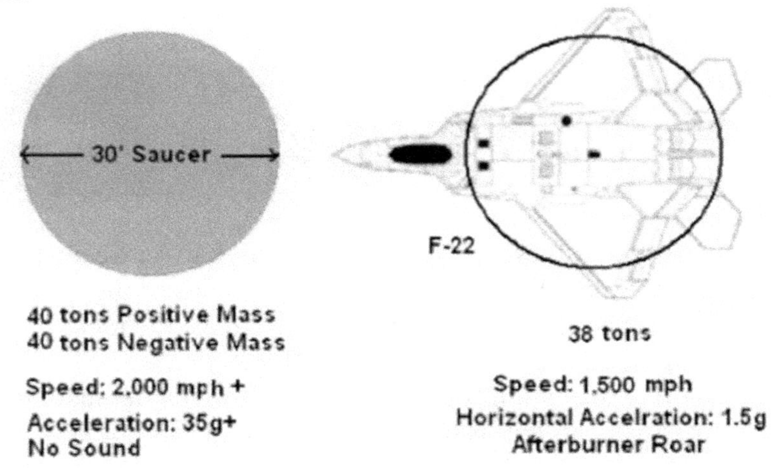

The closest flying thing we have to that size would be a plane like F-22. A superposition of a circular 30' foot object shows that it about matches the substance of the plane. From eye witness reports saucer's appear to be of metal. Since the plane is also, we will take that the have about the same mass. Fully loaded, an F-22 weighs 38 tons. We will therefore take the (positive) mass of the saucer to be roughly 40 tons.

But a UFO has an equal amount of negative mass to the positive so it is buoyant or gravitationally neutral in the Earth's gravity. Therefore we assume that the negative mass in the UFO is also 40 tons.

We know that the negative vector potential **A(-)** it has to generate is very large to cancel the **A(+)** arising from the universal potential. Therefore that negative mass has to be as large as possible to generate this field. Secondly, it has be spun at great speed, because it is essentially the speed that is responsible for the potential.

What about its geometry, its shape? Well, let for the sake of argument say that negative mass in the form of a disk. The formula for the magnetic dipole which generates this field shows that if you spin a charged disk, the field will be proportional to the FOURTH POWER of the radius of the disk . That means is that if you double the diameter of the disk, the field generated will go up by a factor of SIXTEEN. So wouldn't it behoove one to make the mass a very wide thin disk? Take a look at a common flying saucer. Is not flat and wide, to accommodate a wide disk?

The density of matter is determined by atomic structure based on nuclear physics. We know that transuramic elements, which are somewhat heavier than lead and uranium are possible, but they are unstable and short-lived. I doubt that the UFO engineers can really alter nuclear structure and make material more dense. So for our calculations let us limit ourselves to negative matter of density 10 gm/cc, or a little less than the density of lead (11 gm/cc).

The negative vector potential **A(-)** generated by a moving mass depends on the velocity

$$\mathbf{A} = \int \rho v / r \, dV$$

If you integrate over all the matter in the universe, you get the following relationship: An object moving in space, that is, in the background potential **Φ** will generate a vector potential

$$\mathbf{A} = \mathbf{velocity/G}$$

where the **G** is the gravitational constant. This result comes from Sciama (1) in the **ARTICLE** on "Fundamental Equation of a Flying Saucer."

Note that **A(+)** is velocity dependent. Suppose we pick a simple number: velocity = 1 km/sec or 1000 m/sec. This velocity corresponds to 2,200 miles per hour. From the Belgian radar data, such speeds MAY have been recorded for UFOs. Suppose the UFO would like to be free of inertia while traveling at such speeds. The vector potential **A** at this speed would be

$$\mathbf{A(+) = 1000\ m/sec\ /\ 6.6 \times 10^{-11}}$$
$$\mathbf{= 1.5 \times 10^{13}\ kg\ sec/m^2}$$

Any object moving at this speed would be subject to such a vector potential, which, as we have said, would not have any effect. **A(+)** does not create a gravitational field, only **∂A(+)/∂t** does.

Therefore the gravitational engine of our saucer would have to generate a vector potential of **1.5×10^{13} kg sec/meter^2** to to create a negative **A(-)** hole in which to hide, so that variations in **A(+), ∂A/∂t**, do not reach it, that is, be able to overcome inertia when it traveling up to 2,200 miles per hour.

And let us say we move the mass at half the speed of light, **$1.5 \times (10)^8$ m/s.** The relativistic mass increase at this speed, **$1/\sqrt{(1-(v/c)^2)}$**, is **15%.**

For a disk the formula for is:

$$\mathbf{A(-) = \frac{\pi^2}{2}\rho\frac{fa^4d}{r^2}}$$

where **a** the radius, **f** the frequency, **ρ** the density, and **d** the thickness.

Rotating 40 ton Disk

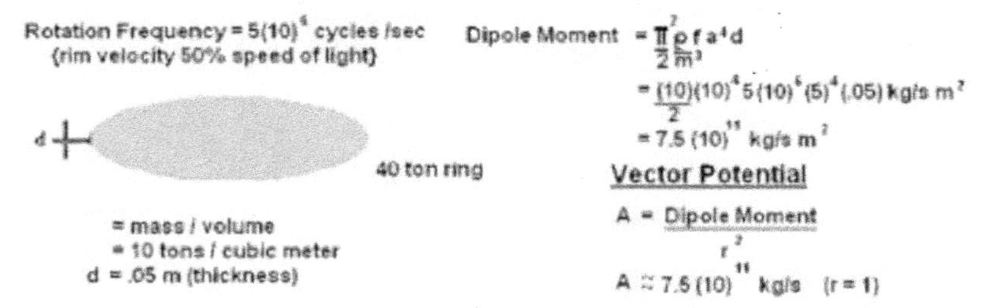

If we spin a 40 ton disk at the rim velocity of half the speed of light, we get only **$7.5 \times (10)^{11}$.** That means the saucer could block inertia only traveling at a very slow speed. At appears that we need to swing the entire mass at that speed. The inner part of the disk travels slower, and does not contribute sufficiently to vector potential.

The formula for the vector potential from a mass current loop is

$$A(-) = \kappa\pi M f a^2/r^2$$

where **M** is the mass of the ring, and **κ** has the numerical value 1.

Rotating 40 ton Ring

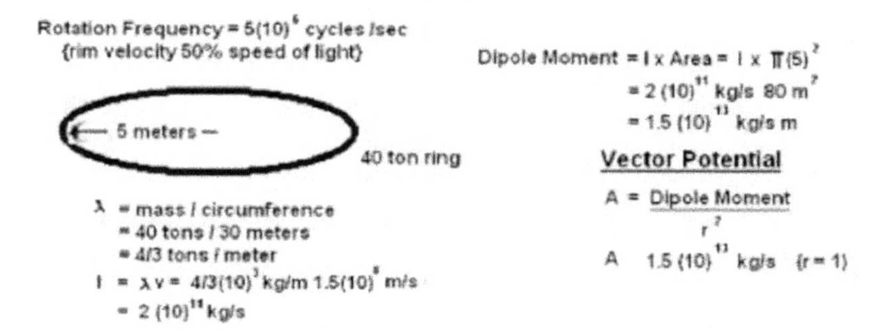

It is only if we concentrate the entire mass in a ring, all at 5 meters from the center, that we can achieve a vector potential of **$1.5 \times (10)^{13}$. kg sec/m^2**

What this shows us, is that if we take into all the requirements a UFO must meet, including densities allowed by nature, the strength of the vector potential creating inertia, and the laws of Special Relativity, not only does it show us that a 30 foot diameter saucer is the smallest possible, but it also determines the geometry of the gravitational engine. And the shape of the engine, a 30 foot diameter negative mass ring in the form of a frisbee, essentially determines the shape the saucer UFO must take to accommodate the engine.

We examine in detail what happens to the ring when it is spun up in Appendix B.

10 The Shape of a Flying Saucer

If you think that flying saucers have a funny shape, the nature of its gravitational engine and its purpose might give some insight why they look the way they do.

The saucer shape of the flying saucer comes from the fact that it has a wide circular flange to accommodate a large rotating ring to create the necessary vector potential **A(-).** The bulk of the positive mass of the saucer is in the ball shaped structure in the center, which is concentrated be inside a circle in an inertia free zone.

The smallness of the saucer limits the amount of negative mass it can use to generate a negative vector potential **A(-)** and the 5 meter radius means it has be spun very fast, at a speed that cannot exceed the speed of light. So we have our work cut out for us.

We set ourselves the task of finding out whether the smallest possible flying saucer, about 30 feet in diameter from eye witness accounts, can achieve ALL the strange behaviors attributed UFOs. This may not be the case. It is possible that some things only larger UFOs can do. But let's try it anyway. If a small flying saucer can do it, then we have settled that ALL UFOs can.

We have shown in the **ARTICLE** that the vector potential such a small saucer can generate, $\mathbf{A(-) = 1.5(10)^{13}\ kg\ sec/m^2}$, is enough to lift a car off the ground at 20 feet while moving at 20 meters per second. We have also shown that when the saucer slows down, the anti-gravity bubble spreads to a mile and a half with no inertia, so the same vector potential can cause headlights and engines to die and for some people to experience paralysis.

The only thing we need to show is the saucer will escape feeling inertia even at 1 km/sec or 2,200 miles per hour. We picked 1.000 meters/second as a nice round number, and because Belgian radar data indicated that UFOs MAY attain speeds of 2,000 mph (see Appendix at page 221). And here it gets difficult.

We have several problems: one is that the simple mathematical calculation of the vector potential **A(-)** breaks down near the saucer. And the other is that we really do not know how the combination of positive and negative mass in the saucer will effect inertia.

The rotating negative mass ring of the flying saucer's engine creates a gravitational magnetic dipole. (The lines in the illustration are gravitomagnetic **B(-)** field lines from which the nature of field is easier to see. The vector potential **A(-)** field lines are perpendicular to these, forming a torus.)

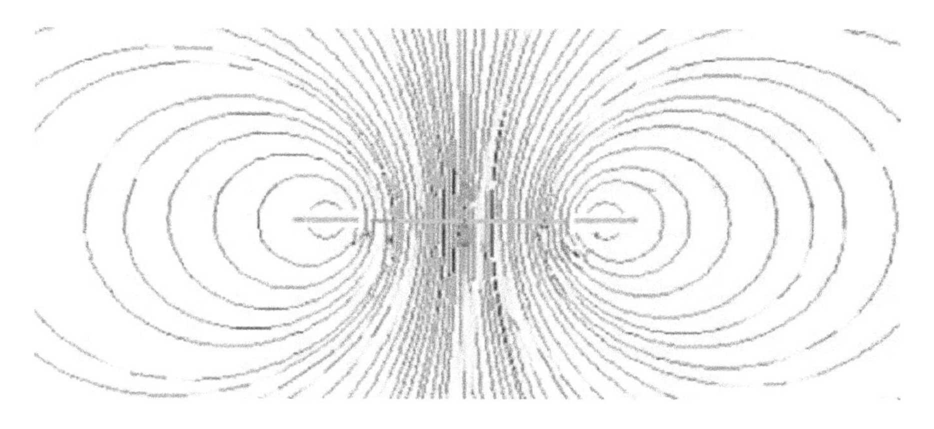

As you see the dipole is an extended object. The intensity of the field lines is greatest inside the ring.

Unfortunately, the formula for the vector potential **A(-)**

$$\mathbf{A(-) = (dipole\ moment)/r^2}$$

is an approximation that is accurate only at some distance from the ring. The formula assumes that the dipole is a point source and there is a singularity at $\mathbf{r = 0}$. Of course the real dipole is an extended object. There is no singularity $\mathbf{r = 0}$.
The actual formula for **A(-)** near the ring involves elliptic

integrals and is complicated. And for an extended Frisbee type ring the radius **a** of the ring has to be integrated from **a = 4** to **a = 5 meters**. We have examined the elliptic integrals and also a power series expansion for **A(-)** near the ring, and it appears that it is reasonably good approximation to take **r = 1** and just to apply the above dipole formula near the ring. That means **A(-)** just equals dipole moment (times some units) at the ring. Whatever error there is appears to be minimal, less than an order of magnitude. An order of magnitude is a good approximation, but it is not good enough for a precise calculation.

The situation we than have is the following: The positive mass of the flying saucer appears to be an approximately spherical structure in the middle of the craft, at the center of the ring where the vector **A(-)** is most intense. We have shown (see also the **ARTICLE**), that radius of the boundary between the **A(+)** and **A(-)** depends on the velocity (see **ARTICLE**).

$$\mathbf{r} \propto 1/\sqrt{\mathbf{v}}$$

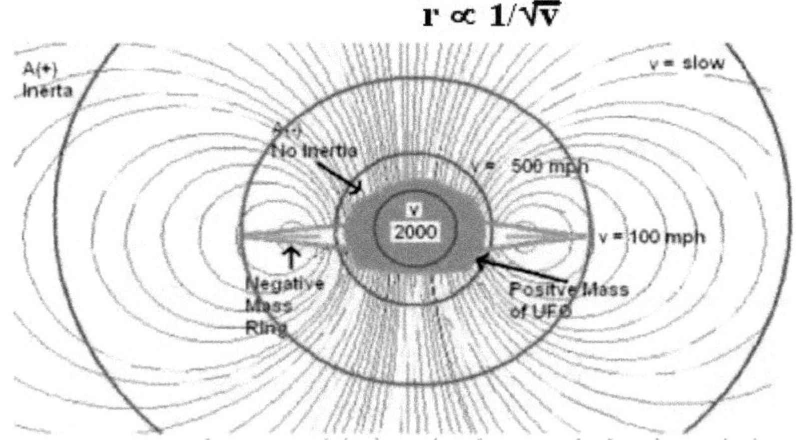

Suppose at some slower speed the boundary between the inertia producing **A(+)** and the **A(-)** is at the outer circle. Everything inside the circle is in the A(-) zone, is inertia free.

At a speed of 100 mph, the entire flying saucer is still within the inertia free zone. The boundary between the positive **A(+)** and the negative **A(-),** indicating a inertia free region, has shrunk to the circle just enclosing the saucer.

If we now raise the speed to 500 mph then boundary between the **A(+)** and the **A(-)** has encloses the body of the saucer, but not the ring. What we notice is that still very little of the flying saucer is subject to inertia. That is, on the **A(+)** side of the circle is the negative mass ring with a covering. But negative mass when immersed in **Φ** creates a "perverse" inertia, an inertia opposite in sign to normal inertia, that actually PUSHES the object, not retarding it. Therefore in the region just outside the 500 mph circle the negative mass rings "perverse" inertia which may cancel out the positive inertia from the positive mass covering of the ring. The bulk of the positive mass structure of the flying saucer is within the small red circle, or inertia free zone.

When the speed increases to 2000 miles per hour, boundary has shrunk to the innermost circle. Here we are not sure what happens. Now all of the negative mass and much of the positive mass of the saucer is exposed to the positive vector potential, to **A(+),** which produces inertia.

At the inner circle the **A(+) / A(-)** boundary is at radius of one meter. If we wanted to the boundary at a 3 meter radius, enclosing most of the positive mass of the UFO, we would have to increase **A(-)** tenfold to **A(-) =** $\mathbf{1.5(10)^{14}}$ **kg sec/m**$^{\mathbf{2}}$ by increasing the spun mass by tenfold, or the rotation speed by tenfold, neither of which is possible.

If you look at pictures of saucers, some have wide cores that appear to be of 3 meter radius, or 20 feet wide. But some have a small core, only one third the diameter of say a 30 foot saucer, that is **10 feet**. If we do the calculation for such a saucer, what we get is that the inertia if free region need only be of **radius = 1.5 meters**, then **velocities of 1,110 mph** would be possible. (The sample speeds reported in the Appendix go up to 1010 knots.) So even in our calculation a 10 feet central core saucer would be inertia free at the highest UFO speed actually recorded.

Could such a saucer accommodate the positive mass which also has to be 40 tons also? The answer is yes. The density of steel is **8 g/cc**. If the core were spherical, it would have a **volume of 14 cubic meters**. The 40 tons of steel would take up only **5 cubic meters** or one third of the volume of the spherical central core.

It is possible that small flying saucer are inertia free only at a lower speed. Our 10 foot core saucer is up to 1100 mph. The only hard data we have for higher speeds, in the possible 2000 mph range, are for Big Black Triangle type UFOs. They most likely have multiple negative mass rings and making the inertia free requirements easier to meet.

I have a feeling that the UFO engineers with their superior understanding and experience are able to squeeze better performance out of their design. The funny shape of the saucer is probably the result of their maximization. I am not smart enough or know enough to figure out how they did it.

And why is the flying saucer round? Why does it not have a front and a back? We will see in the chapter "UFO Anti-Gravity Propulsion" that the UFO is accelerated by a shift of the rotating ring. The acceleration is in the direction of the ring shift. There does not appear to be a mechanism in the UFO that can make it do curved motion. So if it wants to go in another direction, the ring is simply shifted in that direction. The observed motion of UFOs is a series of zig-zags. Therefore the flying saucer is round so it does not make any difference in which direction it wants to go, all direction are equal.

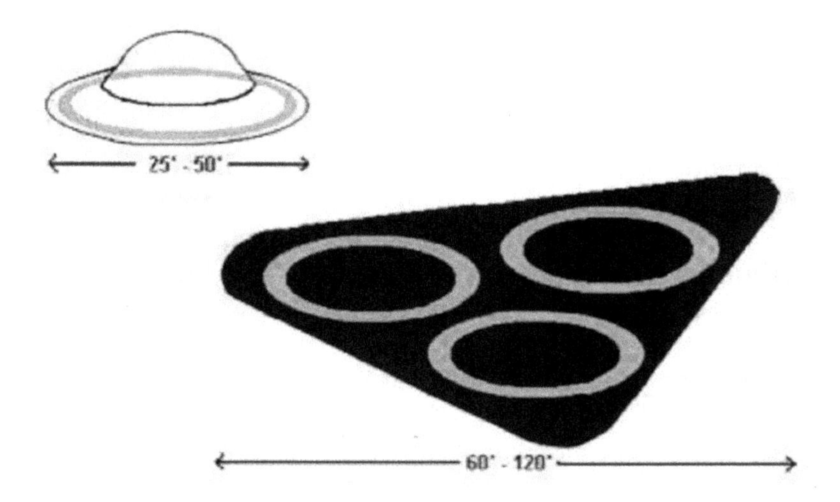

If we take the shape of other larger UFOs, the next in size is the Big Black Triangle. It does not take a great deal of imagination to see that this shape just happens to accommodate three of the rings of saucer side by side.

Of course larger UFOs can have any number of "engines" and their rings can be distributed in various patterns. Here are some sketches

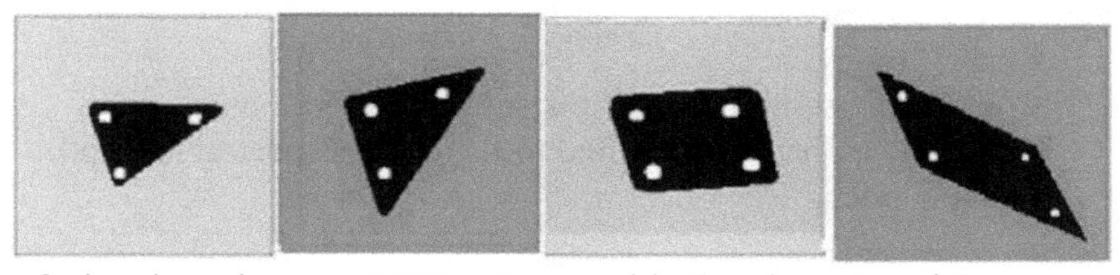

of triangle and square UFOs. Presumably they have 3 and 4 rings side by side. It is estimated that the "Phoenix Lights" UFO was some 3,000 feet in diameter and would have many "engines". (Make a note of the strategically places "lights". We will have an explanation later.)

11 Why Do Some Flying Saucer Glow?

In many eye witness reports UFOs glow brightly, sometimes with an orange glow. The ***UFOCAT*** lists 4251 instances. In one reported mentioned in Dr. Hyneks The UFO Experience. (1):

".....when the UFO landed it changed from the original red-orange color to a bluish green but that when it rose it changed back to the red-orange."

Now when you heat on object to incandescence you can get various shades of yellow, red or white, but never green, no matter what the material you are heating is. Therefore the glow must have been something else. The atmosphere is 71% nitrogen, and it just so happens that nitrogen has strong spectral lines in orange and green.

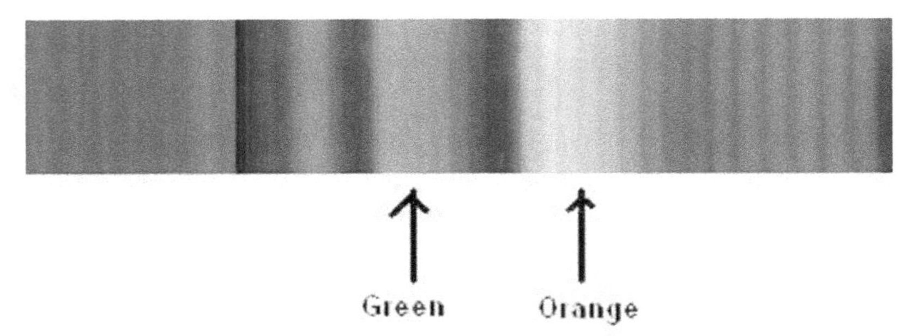

So the UFO glows because it is surrounded by a nitrogen plasma which has been excited to the orange spectral line and then the green.

The orange color of the UFO can be seen in the laboratory where it produced by inductive coupling in a device called the GEC reference cell.

In the cell, microwave energy is introduced by a coil in the cell, which produces an induced oscillating voltage called the EI. This strips the electrons and causes a discharge, and that then emits the glow.

This glow is usually not possible at atmospheric pressure (that is around the UFO) because the electrons immediately bounce off each other at random angles and distributing the induced energy in many frequencies and you do not get a the excitation of a single line like the orange and the green. In the cell this is avoided by keeping the nitrogen gas at a pressure of about 1/1000 of atmospheric so electrons rarely collide.

If one had, however, a powerful static magnetic field in the neighborhood, the electrons would be bound to spiral only along the field lines and not be able to collide at random angles. This is called magnetic trapping. Then the excitation of the electrons would be maintained at the applied EI level, and a single spectral line, like the orange could be excited.

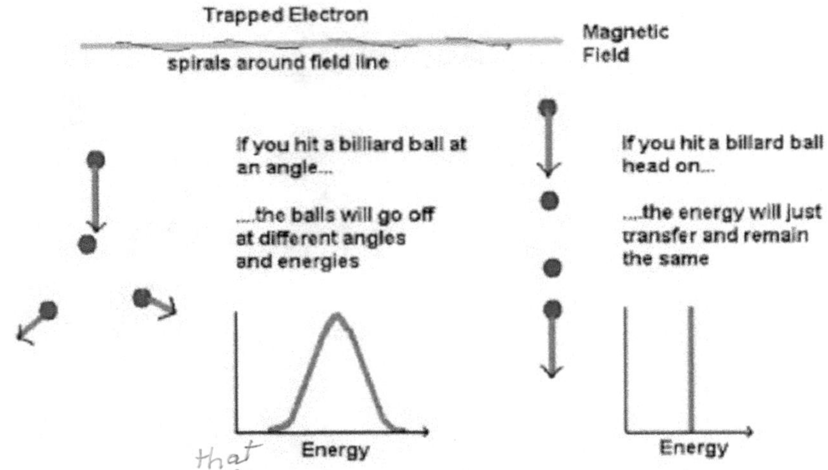

There are magnets that are in the UFO to support the negative matter ring in the UFO in a way that it can move and spin. (For about $40 you can buy yourself a globe that will happily float magnetically on your desk.)

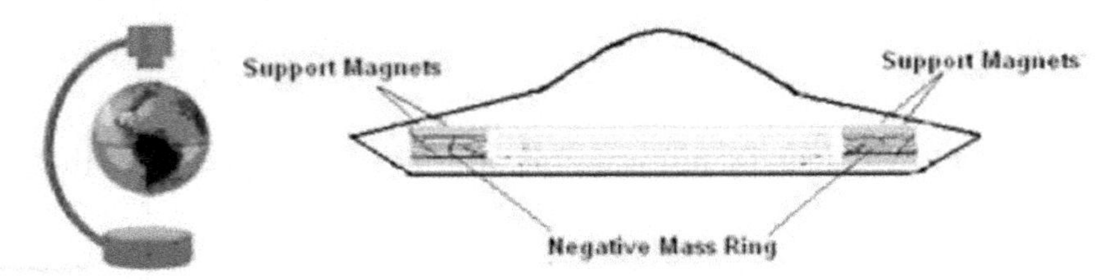

Let's estimate the size of such magnets. Maglev (magnetically levitated) trains they use 5 tesla magnets (50,000 gauss) to support the cars weighing dozens of tons. So let us assume (very conservatively) that two such magnets support the ring, one above the other below, or 10 tesla all together.

There are 150 reports in ***UFOCAT*** that airplane compass near UFOs being thrown off. The earth's magnetic field is about .3 gauss. Let us suppose a compass can be thrown off by a stray .1 gauss field.

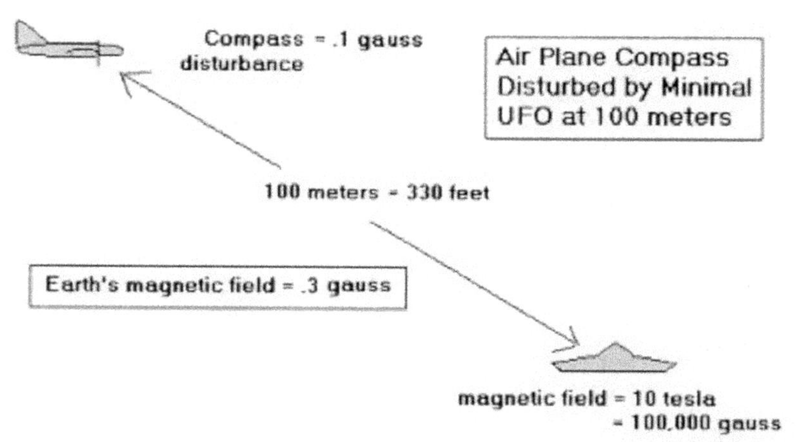

How far could the effects of such magnets reach? It turns out that the compass in an airplane 100 meters or 330 feet, or slightly more than a football field, will be disturbed. There is therefore evidence based on the aircraft reports that UFOs have large static magnets of 10 tesla or more.

But the plasma in the GEC cell was driven by RF radiation in a coil. Is there such in the UFO?

There are 381 reports in ***UFOCAT*** of disturbance of radio and TV reception near UFOs. That indicates that the UFOs are putting out considerable RF radiation.

A wide range of RF frequencies can excite a plasma in a GEC cell, anywhere from one kilocycle to 100 megacycles. The standard frequency for the cell is 13.4 megacycles. As to the disturbance in radio reception, the AM band is around 1 megacycle, so that such a frequency from the UFO could easily excite the plasma and produce the orange of green glow, just as in the GEC cell – and create radio and TV disturbance

But the glow exists even when the UFO is stationary on the ground. Therefore whatever the UFO uses the RF energy for, it is not propulsion.

As to the change in color upon landing: Suppose we consider an equivalent circuit how the RF induced voltage in the GEC cell works.

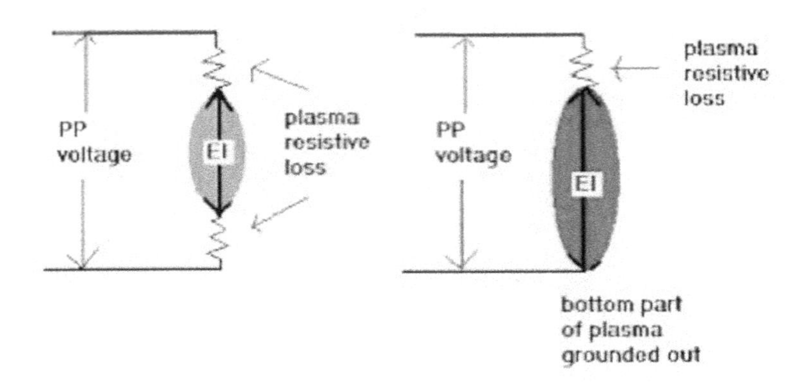

The circuitry of the cell puts out voltage called the Point to Point voltage. Of this some is lost in the resistance of the plasma so only part of it is the EI which excites the atoms and the spectral line. The energy of the green line is higher than that of the orange line, therefore more voltage would be necessary to excite the green rather the orange.

As the UFO lands, the bottom part of the plasma surrounding it touches the ground. The ground acts like a grounding plate short circuiting that part of the plasma. If some part of the plasma this is shorted out, then more of the point to point voltage is available for excitation and the EI would be higher, and this higher voltage could excite the green line. When UFO takes off, the shorting by the ground plane disappears, and the color turns orange again.

What does the UFO use the RF radiation for, if not for propulsion?

In maglev trains, tracks coils are embedded in the track at close intervals. Alternating current is sent through these coils. The coils induce currents in the maglev car which then interact with the coils and this pulls the car forward.

If you were to wrap the coils around, say, a metal ring, feed alternating current to the magnets , the ring could be made to move, to spin. Indeed this is principle of the alternating current induction or motor .

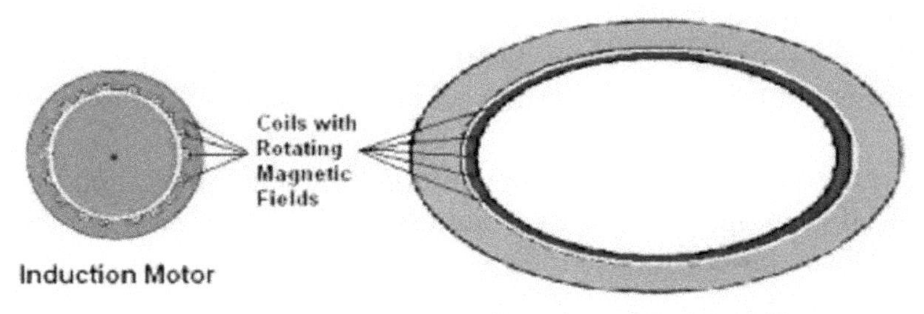

The alternating current frequencies in the maglev coils or the induction motor coils are in the hundreds of cycles per second. If you need to spin a large ring very fast, you would use powerful RF fields in the megacycle range, and this is what UFOs do.

The ring they spin is the gravitationally negative mass of the UFO. The spinning ring creates negative gravitational field which repels the positive gravitational potential of the universe. It creates a bubble in which there is no inertia

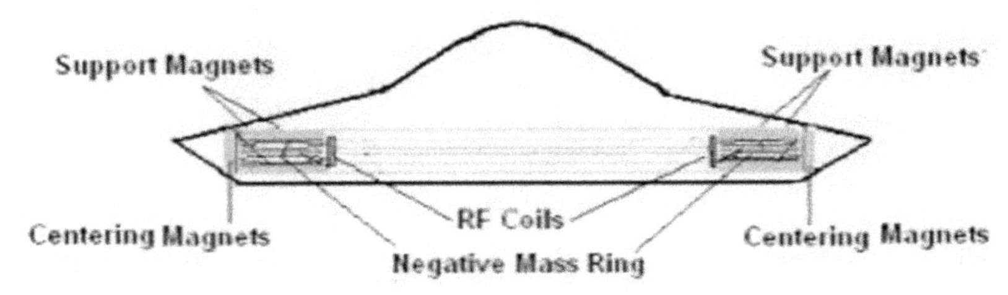

So it is the powerful magnets that support the negative mass ring, and the RF radiation which spins the ring, that excite the nitrogen plasma around the UFO, and that makes the UFO glow.

But if all UFOs are inertia free, why do only some flying saucers glow? In particular the big Black triangle UFOs are quite dark.

The big black triangles all have many lights, however. In particular they have a number of white lights about one meter in diameter that cast no beam. The Belgian eye witnesses universally reported no "bascule" or beam from these "lights". What kind of light casts no beam? The answer is when is the light is not a light but a plasma port

If you take a look at the famous Petit-Rechain photograph of the UFO, you will notice the light from the three outer sources is defuse and not focused.

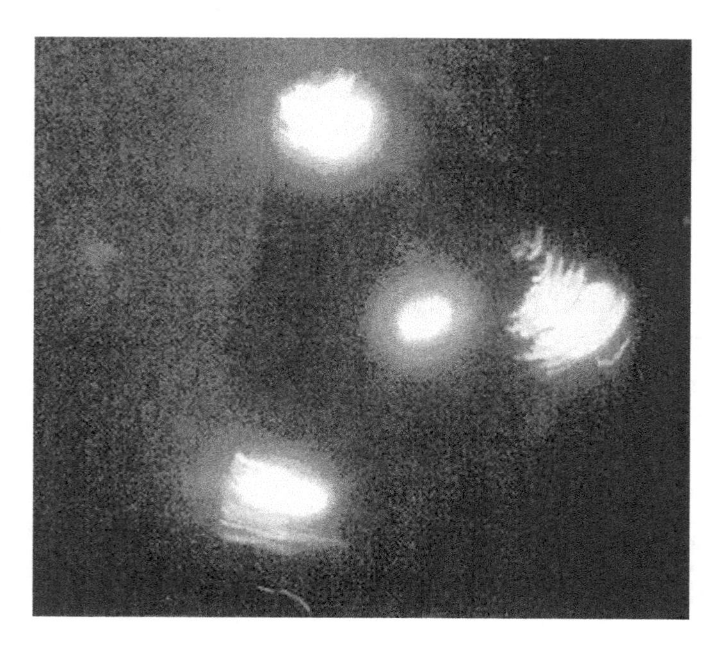

The glow of the flying saucer has been replaced by the glow of these plasma ports. They dispersed the energy from the spinning ring or rings. They in effect act like radiators in a car. The **Second Law of Thermodynamics** mandating the dispersal of energy is obeyed.

12 Robert Forward's Anti-Gravity Propulsion

UFOs do not interact with the environment when they propel themselves. They make no sound, create no turbulence, or leave a wake. They have no traction, that is, they do not pull or push against anything physical to make themselves move. The question is, do they use some form of what is called anti-gravity propulsion?

In 1957 Herman Bondi wrote a paper (1) that showed that gravitationally negative matter was not contrary to General Relativity. In the paper he says:

"Imagine a body of positive mass and negative mass separated by empty space. Then, to use the language of the Newtonian approximation, the positive mass body will attract the negative one (since all bodies are attracted by it), while the negative body will repel the positive body (since all bodies are repelled by it). If the motion is confined to the line of centers, then one would expect the pair to move off with uniform acceleration."

This broached the notion of anti-matter propulsion.

The idea is very simple. Suppose you throw a baseball. As you make the throw, the ball pushes back against your hand. This push is force of inertia. Now suppose the mass of the basball was negative. Then **F= - ma**, the force of inertia is in the opposite directions. The baseball actually goes flying out of your hand in the forward direction.

This notion was further explored by Robert Forward in 1988 in a paper (2) and a report to the AIAA/ASME/SAE/ASEE Joint Propulsion Conference (3). Unlike Bondi, who considered connection between masses only by gravity, Forward also explored connection by a spring, electric forces, etc.

In the paper Forward describes how anti-matter propulsion might work:

"To make an "ideal" propulsion system, we just need a negative matter particle in the "engine" room that that has a mass that equals in magnitude the positive mass of the entire ship. If we want to go forward, we pull a spring from the back wall of the engine room and hook it to the negative matter particle. Immediately, the perverse inertial reaction of the negative matter particle will cause it to accelerate off in the forward direction , pulling the spacecraft forward with an acceleration that is proportional to the strength of the spring. To stop accelerating, merely unhook the spring. To decelerate the ship to a stop, replace the spring coming from the back wall of the engine room with a spring coming from the forward wall. Simply, isn't it?"

Here the force of gravity is replaced by the force of stretching of a spring. Say the positive mass he envisions is on the left and the negative on the right. Stretching a spring between them, the negative mass is pulled to the left. But because the mass is negative (remember $F = - ma$) the negative mass actually accelerates to the right. And because the masses are connected by the spring, they both move off together to the right with constant acceleration . This is also what Bondi said. This is the basic idea of negative-mass acceleration.

From his analyses, Forward and comes to conclusion that in no case he considered, no matter how the masses are connected, are the laws of conservation of energy or conservation of momentum violated.

If we replace Forward's spring by the magnetic suspension of the negative mass which is likely in UFOs, the suspension having a spring constant called the magnetic "stiffness", then his description might be applied to UFOs.

We will now investigate if UFOs actually use Forward's anti-gravitational propulsion method.

In essence, anti-gravity propulsion would work like this: all one need to do is to shift the center of mass of the negative mass relative to the center of mass of the spacecraft, and the spacecraft will follow. The force for the motion would be provided for the magnetic "spring" which is compressed by the shift. The force then generates a "perverse" acceleration which moves the system.

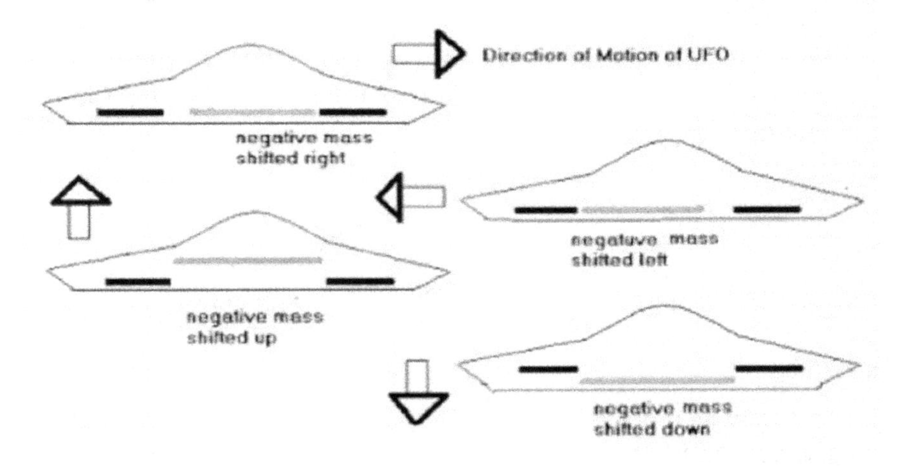

As long as there is a shift, there will be acceleration. To stop the acceleration and just coast, one simply centers the ring. In practice one pulses a displacement for a short time to accelerate, then the motion at the attained velocity will continue.

We have abundant eyewitness data that flying saucers cause disturbances in the compasses of nearby aircraft (150 reports in ***UFOCAT***). Working back from that data one gets that saucers have large stationary magnets of at least 10 tesla. Such magnets are used in the elevation of maglev trains and are strong enough to act as bearing for the large negative mass rings in the UFO.

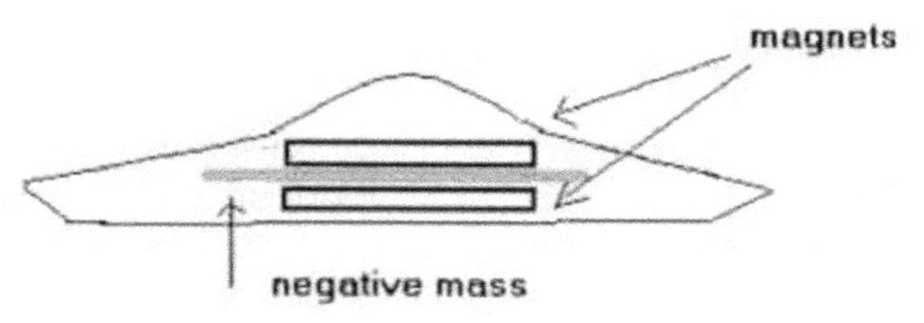

There are also hundreds of reports of radio and TV disturbances near flying saucers (385 reports in ***UFOCAT***). That is an indication that the spacecraft are a source of powerful microwave radiation. These RF fields can be used by the UFO to spin the ring at a high (megacycle rate of spin) which is necessary to create the negative fields to counteract inertia. Very briefly, we conclude that the situation might look like this:

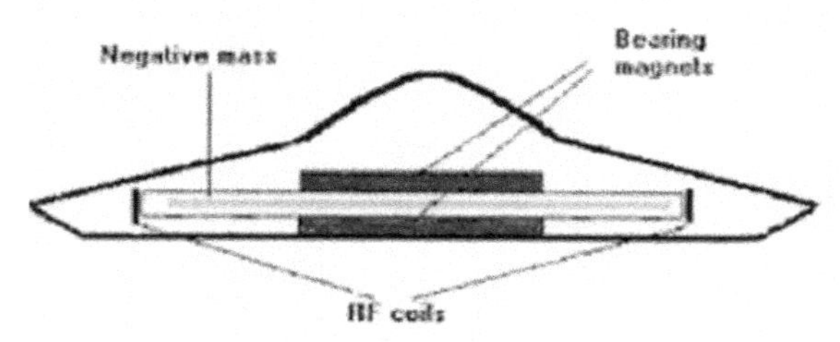

As long as there is a shift, there will be acceleration. To stop the acceleration and just coast, one simply centers the disk removing the pressure off the magnetic "spring'. In practice one pulses a displacement for a short time to accelerate, then the motion at the attained velocity will continue.

What you might notice is that the motion of the ring for displacement is extremely confined, both by the position of the magnets, and the driving coils. The question we have to answer is, is it possible to create anti-gravity propulsion as envisaged by Forward which such limited movement ?

According to Forward, the force of the spring creates the acceleration: **kr = aM**
where k is the spring constant, r stretch of the spring, a the acceleration, and M the mass of the spacecraft. We get an idea of how strong the magnetic forces have to be from the fact

magnets act as bearings to support spinning anti-matter ring. We have found (see **ARTICLE**) the mass of the ring is upwards of 40 tons. Lets choose 50 tons for illustration. Let us also assume that the saucer can do an upwards acceleration of 100 meters/sec/sec, or 10 times terrestrial gravity. The left side of the equation then yields

$$\mathbf{kr = (100\ m/s/s)(50\ tons)}$$
$$\mathbf{= (10^2\ m/s/s)5(10^4\ kg)}$$
$$\mathbf{= 5(10^6)\ newtons,}$$

To find the displacement, **r**, needed we need to get an idea of the stiffness of the magnets.

In the book ***Superconducting Levitation*** (4), on page 137 there is a graph of stiffness vs. levitation force obtained by experiments on a YOCO super conducting coil. In the text they state a formula derived from the graph agrees very well with experiment.

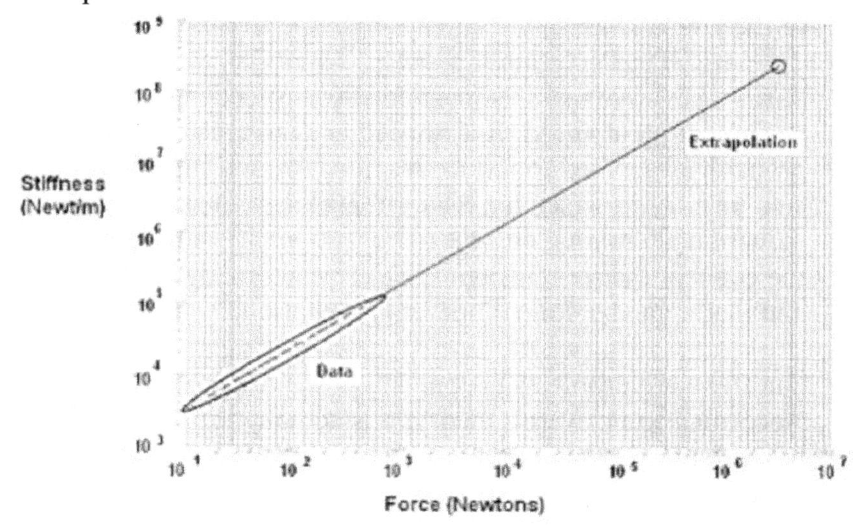

We now extrapolate with the results to the region of $5(10^6)$ newtons. The stiffness at this levitation force seems to be $5(10^8)$ newtons/meter.

Then the displacement necessary to achieve the 100 m/s/s acceleration is

r = <u>5(10⁶) newtons</u>

$$5(10^{8})\ \text{n/m}$$
$$= (10^{-2})\ \text{meters}$$
$$= 1\ \text{cm.}$$

That means that an exceedingly small displacements of the spinning ring will produce appreciable acceleration. This may be surprising. But remember in Forward's theory it is the force of the spring which creates the acceleration. The suspension magnets are extremely stiff because they have to support the tens of tons of weight of the craft. A small movement against this substantial force produces the acceleration.

We can now see how UFOs can make high acceleration motions without using energy for propulsion directly. All they need to do is to displace the spinning negative mass slightly, and the craft will instantaneously follow. There is no interaction with the environment, the atmosphere. There are no sounds, turbulence, or light. The propulsion is totally internal to the UFO.

A large negative mass is magnetically suspended to allow it to be spun at great speed. It is housed in a large can which is probably evacuated to reduce friction. Around the edges of the can are coils to which the RF current is sent to create an induction alternating current motor. So what we have here is the concept that propulsion of UFOs is internally generated, and that it can be done if one has gravitationally negative mass.

But there are problems in applying these ideas to UFOs. According to Forward's theory, it is actually the force applied to the "spring" which through Newton's Second Law generates the acceleration, then experienced by both positive and negative masses as they are connected. But we have seen accelerations of UFOs of 22g and 35g. The **35g** acceleration would require a 3.5 cm displacement of the negative mass for a 50 ton UFO, this is not a problem. The problem is that such accelerations would require the application of some **17.5 million Newtons** , or **1,750 tons** of force, the weight of a small ship. This amount of force is prohibitively high to be generated in a small saucer. The apparatus involved in the shift would simply not have the material strength to withstand such loads. Therefore Forward's anti-gravity propulsion cannot be the source of motion in UFOs.

The problem is that Forward's theory considers only the "perverse" inertia of negative mass. Both Bondi and Forward presumed that their masses were in ordinary space, that is immersed in universal background potential. The propulsion they envisaged would come from the "perverse" inertial acceleration of the negative mass, the inertial force being derived from the huge universal background scalar potential. But we know that in a UFO the effect of this scalar potential is neutralized. There most be another source for the force of propulsion.

The actual force, the power of the anti-gravity drives comes from the negative vector potential generated by the spinning negative mass ring. We have already seen this to be real powerful force. A nearby passing UFO can lift a car. It is the displacement of the ring with respect to the positive mass structure of the UFO that generates the gravitational force which propels the UFO.

And the reason that the tremendous forces required by Forward are not needed, is that in the negative gravity bubble of the UFO the inertial mass of the UFO is reduced by a factor of a million, allowing smaller forces to generate the tremendous accelerations observed.

This is a case where the back of the envelope calculations mentioned earlier lead to an unacceptable result. Further reflection then revealed the proper solution to the problem.

Also both Bondi and Forward consider constantly accelerating systems. This does not square well with eye witness reports. They indicate that the UFOs tend to do accelerations in bursts, and then coast at constant speed. This was remarked on specifically by the analysis of Nellis data.

Robert Forward himself was dissatisfied with his anti-gravity propulsion theory. His physicist's instinct told him something was just not right. At the end of his paper he has a "Personal Note to Skeptical Readers".

"Look, I know that negative matter propulsion is ridiculous and logically impossible. If so, it should be easy to prove it is logically impossible. But Bondi could not do it. Hoffman could not do it. Terlestki could not do it, Bonner could not do it, and I could not do it. Can you?"(4).

We will explain in detail later how UFO propulsion works. Let us for the time being just adopt the idea that the shifting of the negative mass of the in the UFO with respect to the positive generates the anti-gravity drive.

1.Bondi, H. Negative Mass in General Relativity, Reviews of Modern Physics, Vol. 29, No.3, July 1957, pp. 423-428.
2. Forward, R. L., Negative Matter Propulsion, J. Propulsion, VOL 6, No. 1, p. 38 (1988)
3. Forward ,R. L., "Negative Matter Propulsion", AIAA Paper 88- 3618, July 1988.
4. R. L. Forward, Negative Matter Propulsion, J. Propulsion, VOL
6, No. 1, p. 38 (1988)

13 Kinematics of the Upward Falling Leaf Maneuver

The ***UFOCAT*** documents 180 instances of pendulum or "falling leaf" maneuver by UFOs. In two of those instances (Reports Number 16,982 and 87,715) the maneuver is executed in reverse. The UFO falls like a leaf UPWARDS before accelerating away. UFOs are perfectly capable of going straight up or down. Belgian radar recorded a drop by a UFO of 3,000 feet in 2 seconds. So why the occasional pendulum or falling leaf motion?

There do not appear to be any sketches by witnesses of this motion, upwards or downwards, but it probably looked something like this:

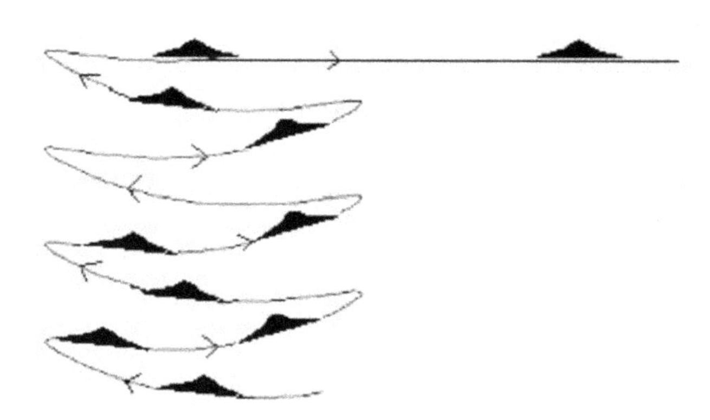

If people use descriptions like "pendulum" and "falling leaf", they are trying to compare what they see to motions that they know. This leads one to believe that the period of the motion is in the neighborhood of 2-4 seconds, otherwise they would have used some other metaphor. Some of the swaying was very slow, so probably with a period of several seconds. Others use the falling leaf metaphor even when the UFO was not falling but swaying in place.

But the motion of a pendulum is part of a circle. At small

angles there is practically no difference between circular and parabolic motion:

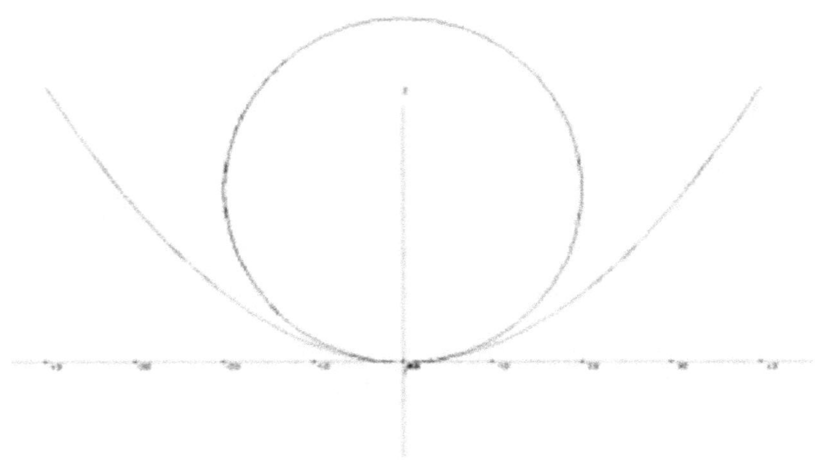

It is highly likely that what they were seeing was parabolic because a parabola is a lissajou figure, and this figure is generated by two perpendicular oscillations, which are more likely for the motion of the negative-matter propulsion of the UFO. The negative matter ring can oscillate in two directions.

For two oscillations to generate a parabola, there are very stringent conditions on the oscillations. The upwards oscillation has to have exactly twice the frequency of the side-to-side, and must be 90 degrees out of phase with it. The equations are:

$$\mathbf{x = \cos(\omega t).}$$
$$\mathbf{z = \cos(2\omega t + \pi/2)}$$

We flatten the parabola by a factor of .1 to get the "falling leaf" look and we add a motion in the z direction, again reducing by .1, to obtain the "upward falling" trajectory in the diagram. The diagram above is generated by the equations

$$\mathbf{x = \cos(\omega t)}$$
$$\mathbf{z = .1\cos(2\omega t + \pi/2) + .1\omega t.}$$

To have such stringent condition on the oscillations is puzzling. Supposedly different spacecraft were exhibiting this behavior and there would be different confinement or restoring forces of the ring in the horizontal and vertical directions in different designs. Something else must be operating to produce this pattern. And it turns out that something else is the well-known instability in the motion of bodies called the "Dutch Roll".

The name comes from fact that when the Dutch speed skaters on the frozen canals pushed off with one skate, putting their weight on the other, their bodies would dip slightly. When the weight transfer was complete and they pushed off on the other foot, there bodies would dip slightly in the other direction, causing a coordinated up and down and side-to-side motion. It turns out that the **"Dutch Roll"** occurs in other moving vehicles, airplanes, trains, etc.

The motion of any moving body is described by six differential equations covering the six degrees of freedom: forward, up-down, left-right; and rotational ones: pitch, yaw, and roll. The same equations apply to all moving bodies. A single set covers all airplanes, with the only difference being the forces generated by various aero- dynamic surfaces like wings, tailplanes, and ailerons which are then entered into the equations.

But UFOs have no aerodynamic surfaces. They generate their propulsion and all the associated motions purely internally. Then how can the "**Dutch Roll**" phenomenon apply to them? It turns out that the magnetic suspension of the spinning ring has a great similarity to magnetic suspension of maglev trains, and maglev trains exhibit **Dutch Roll**.

A maglev train rides on a fixed conducting track. As the train begins to move the magnets in the train (usually superconducting) induce currents in the track by **Lenz's Law**. These currents generate an opposing magnetic field which repulses the field of the train's magnets and lifts the train above the track.

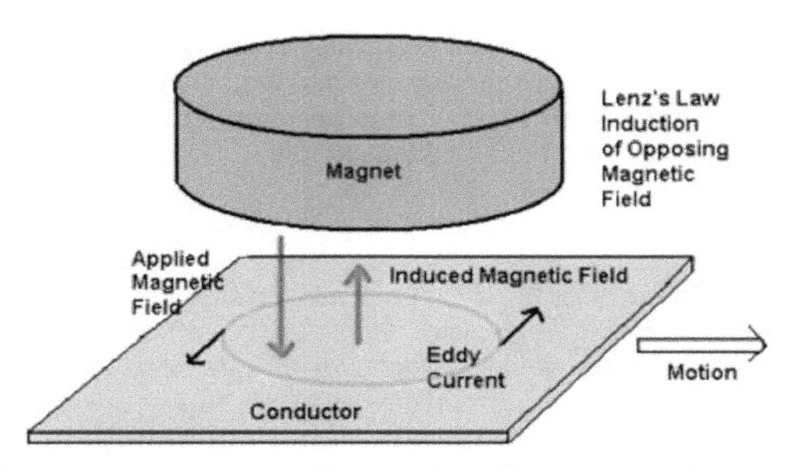

We do not know precisely how the magnetic suspension of the negative matter ring works in the UFO. But let us suppose the ring is a conducting substance like metal. (That would be consistent with it being the rotor in an induction type motor employed for the spin.)

The negative mass counteracts the gravitational pull on the positive mass of the craft. It spins under one of the bearing magnets. The motion of the spin induces currents because of the bearing magnet and the repulsion created actually supports the craft. So the moving maglev train is supported by induced currents in the stationary track, while the stationary is UFO supported by induced currents created by the bearing magnets in the spinning ring.

Cai, et. al. (5) have analyzed a magnetically levitated vehicle. The studied vehicle had the magnets running on L shaped conducting tracks, which provided for lift and also lateral restoring forces.

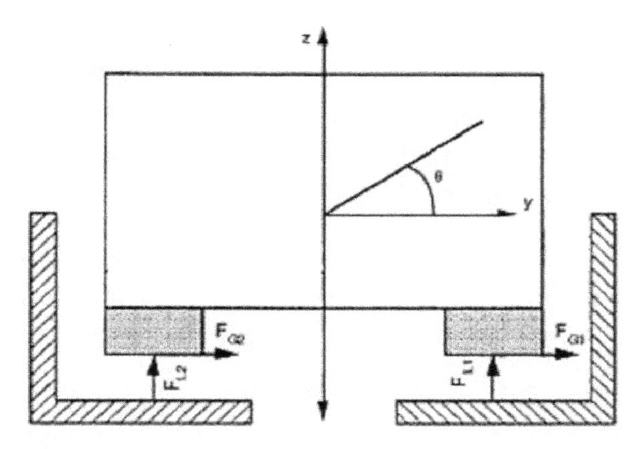

To study instabilities of the vehicles motion, they considered only three of the possible six degrees of freedom. The only force equations they considered where lifting and lateral restoring forces, and three motions: heave, sway and roll.

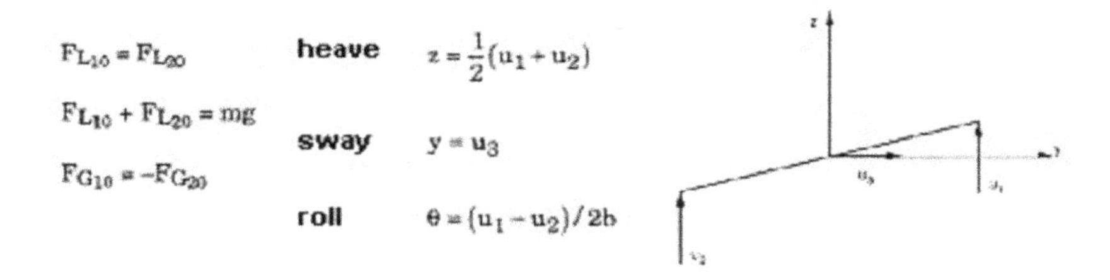

Real eigenvalues correspond to damping and imaginary ones to oscillations The equations usually considered in stability studies are linear first order differential equations which have exponential solutions.

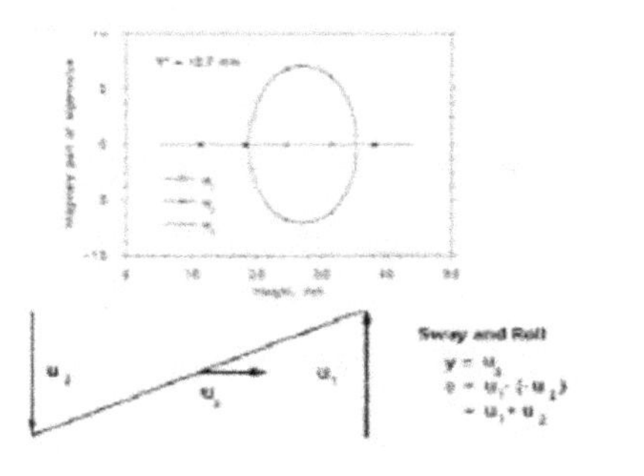

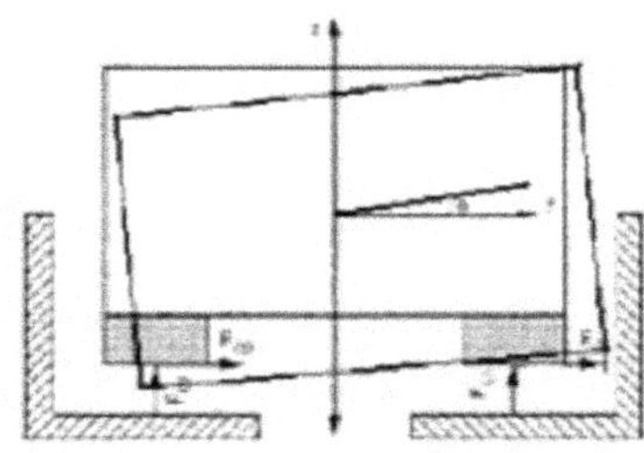

In their analysis the Cai team found modes with imaginary eigenvalues with the modal diagram indicating roll and sway such as in the Dutch Roll mode.

Does this analysis transfer to the anti-gravity drive of a UFO?

We begin with a more realistic arrangement of the magnets and spinning ring. There must be restoring forces to center the ring as these forces are needed to provide anti-gravity propulsion in the horizontal direction. Let us indicate this by placing magnets in a ring around the can of the ring with the coils inside the magnets near the rotor for inductive drive purposes. Also we extend the bearing magnets all the way across the ring so they generate induced **Lenz Effect** currents for support.

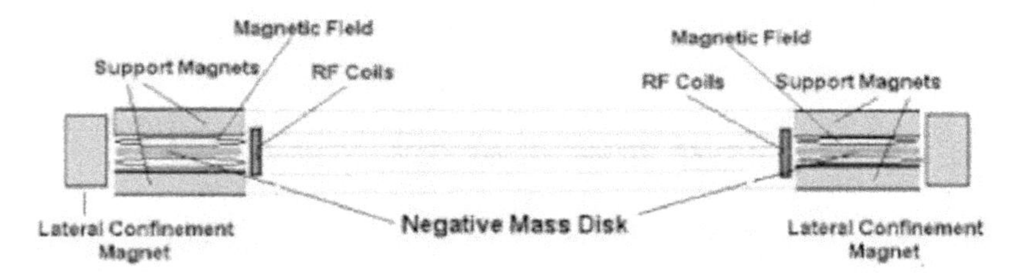

The linear motion at the rim of the ring generates current and provides lift. Therefore this is similar to the maglev case which rides on tracks on each side.

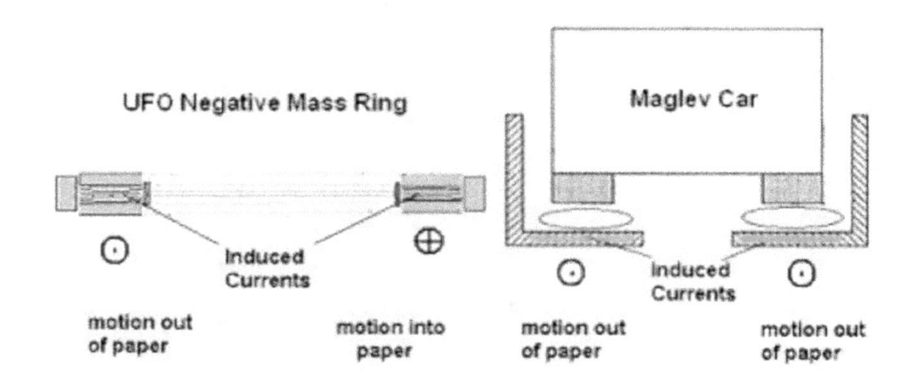

The only difference is that where in the maglev case the current generating motion is linear in one direction (say out of the paper), in the UFO case the motion of the rim is out of the paper one side and into the paper the other side. Does this introduce a difference?

The answer is no, because the instability analysis considers only motion in the lateral plane. The only forces in the equation are **FL**, lifting forces, and **FG**, guiding forces. The forward motion does not appear anywhere in the analysis except to provide lift. And that lift is provided no matter what direction the magnet moves over the track or ring. Therefore the analysis for the maglev vehicle and the UFO supported by the spinning ring appear to be identical.

If the maglev moving over the track exhibits **Dutch Roll**, then the relationship of ring to magnet in the UFO should also exhibit **Dutch Roll**. And since the relative motion of the ring and UFO generate anti-gravity propulsion, this relative **Dutch Roll** motion is mirrored by the UFO and it exhibits the pendulum or "falling leaf" behavior.

The only other thing need for "falling leaf motion" is vertical movement superimposed on the oscillation, either upwards or downwards motion, as we have done in the first diagram.

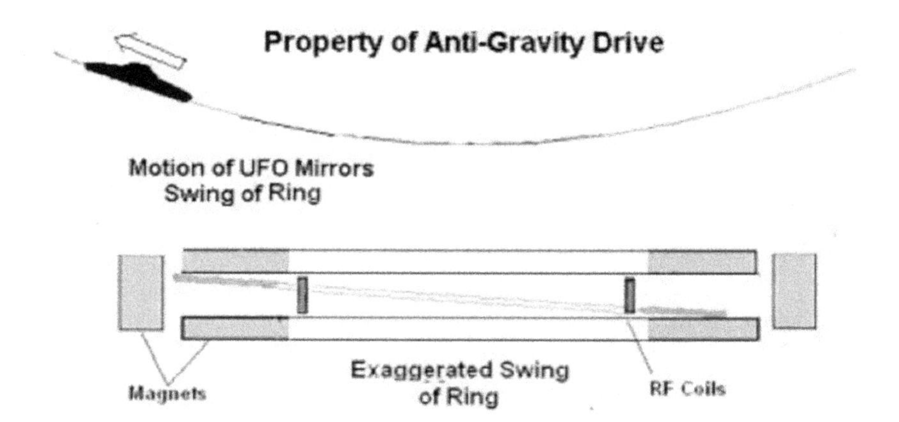

There is no reason why in the UFO vertical and oscillatory motions would not be independent of each other, and so they can be added.

At this point it is not clear why this particular motion of the UFO appears or what triggers it. We know it occurs during vertical ascents and descents and some hover, but not always. The **Dutch Roll** instability in airplanes is always somewhat present, and can be triggered by sharp aileron or rudder pulses. How pronounced it is depends on what steps in the design are taken to dampen it.

It is possible that in UFOs it is a pilot induced oscillation in a careless transition form horizontal to vertical flight. The shifting of the ring upwards or downwards in the anti-gravity drive somehow induces the motion. Because the **Dutch Roll** instability is inherent in magnetic levitation, it might easily be triggered, and in the case of the UFO relatively harmless, it might just be tolerated.

What we have seen is that "upward falling leaf" motion has a ready explanation as an instability in the anti-gravity drive mechanism of the UFO.

It would appear that the existence of this particular motion is fairly convincing evidence that UFOs employ some form of anti-gravity drive, analogous to what was envisioned by Robert Forward 30 years ago.

14 A Dissertation On Mass

There is only one physical mass m. This mass has a positive gravitational charge. And the charge produces a miniscule positive gravitational field. This mass and its charge and field never change. But the miniscule field of this mass interacts with other gravitational fields. This interaction produces gravitational forces. In the equations for these forces the mass appears. If it appears in the equation for Newtonian gravity, we call the mass gravitational mass, $\mathbf{m_g}$.

$$\mathbf{F_g = - m_g G \nabla \varphi}$$

If it appears in the equation for force of inertia, we call it inertial mass, $\mathbf{m_i}$.

$$\mathbf{F_i = - m_i G \frac{\partial \underline{A}}{\partial t}}$$

We can think of these two masses a proportionality constants relating gravitational fields to forces. They are just numbers.

It is well known that Einstein insisted that gravitational mass and inertial mass where identical. It is fundamental to the General Theory. As a matter of physical fact, experimentally the identity of gravitational and inertial mass has been established with a high degree of accuracy which no one really questions.

It turns out that there is a simple physical reason why in our universe inertial and gravitational masses are equal. Gravitational mass $\mathbf{m_g}$ appears in the Newton's gravitational force equation

$$\mathbf{F_g = - m_g G \nabla \varphi}$$

where $\boldsymbol{\varphi}$ is the scalar potential of a body. On the other hand inertial mass, $\mathbf{m_i}$, appears in the force equation for inertia

$$\mathbf{F_i = m_i a = - m_i G \partial A / \partial t}$$

But the acceleration due to inertia comes from the fact that the force is derived from the time derivative of the vector potential, $\mathbf{\partial A / \partial t}$, which $\mathbf{A}$ arises with motion against the universal background potential, $\mathbf{\Phi}$.

$$\mathbf{A = v\Phi/c^2}.$$

$$\frac{\partial \underline{\mathbf{A}}}{\partial \mathbf{t}} = \mathbf{a\Phi/c^2}$$

The $\mathbf{\Phi}$ is merely the accumulations of all the **φ's** in the universe.
Therefore we have for $\mathbf{F_i}$, a being the acceleration **dv/dt**,

$$\mathbf{F_i = - m_i a = - m_i G a \Phi/c^2}$$

And $\mathbf{F_g = - m_g G \nabla \varphi}$

both $\mathbf{F_g}$ and $\mathbf{F_i}$ have the same gravitational potential on the right hand side, $\mathbf{\Phi}$ and the accumulation of all **φ's, Φ.**

Since the gravitational force $\mathbf{F_g}$ arises from $\mathbf{\varphi}$, and the inertial force $\mathbf{F_i}$ arises from the accumulation of the same **φ's**, namely $\mathbf{\Phi}$, it is not surprising that the two masses, $\mathbf{m_g}$ and $\mathbf{m_i}$, are identical. Therefore that the inertial mass and gravitational masses are equal, is true. Einstein was right, as usual - when he was talking of gravity in our Universe.

But Einstein went further. He stated that the identity is fundamental reality upon which General Relativity rests.

Einstein published his Theory of General Relativity almost a hundred years ago, in 1915. It has been rightly celebrated as a great achievement. It has withstood the test of time. Many modifications have been proposed, but none has taken root.

General Relativity is a classical deterministic theory. It is based on Riemann's geometry of curvilinear coordinates. The mathematics is real, i.e. it does not involve complex numbers. In the Theory Einstein recast gravitation into geometry of space. It is a very accurate description of the structure of our universe.

His reasoning was as follows: in curvilinear coordinates, if you want to take the derivative of the vector in Riemannian Geometry, additional terms to the derivative appear because of the nature of curvilinear coordinates.

These are the Γ terms:

$$A^{\mu}_{;\sigma} = \frac{\partial A^{\mu}}{\partial x_{\sigma}} + \Gamma^{\mu}_{\sigma\alpha} A^{\alpha}$$

The equation of geodesic line, which describes linear motion in space is in Riemannian Geometry is

$$\frac{d^2 x_{\mu}}{ds^2} + \Gamma^{\mu}_{\alpha\beta} \frac{dx_{\alpha}}{ds} \frac{dx_{\beta}}{ds} = 0.$$

This is equation of motion in General Relativity. The first term is the acceleration term, and the second term contains gravity.

In Einstein's own words:
"The equations express the influence of inertia and gravitation upon the material particle. The unity of inertia and gravitation is formally expressed by the whole left-hand-side has the character of a tensor (with respect to any transformation of co-ordinates), but the two terms taken separately do not hav tensor character. In analogy with Newton's equations, the the first term would be regarded as the expression for inertia, and the second as the expression for the gravitational force."

He then concludes:
"The possibility of explaining the numerical equality of inertia and gravitation by the unity of their nature gives to the general theory or relativity, according to my conviction, such a superiority over the conceptions of classical mechanics, that all the difficulties encountered must be small in comparison with this progress."

In other words, the necessity of inertial and gravitational mass being equal, follows from the fact that is it necessary for the Theory to fit the curvilinear geometry of Riemann. It is not a statement about the physical property of the masses. It is a requirement to fit a particular mathematical picture.

While we are at it, it easy also to see why the factor $\mathbf{G/c^2}$

appears in Gereral Relativity. In the equation of motion

$$\frac{d^2x_\mu}{ds^2} + \Gamma^\mu_{\alpha\beta}\frac{dx_\alpha}{ds}\frac{dx_\beta}{ds} = 0.$$

If one considers only single source of gravity which is stationary, the whole second term just becomes

$$\frac{d^2x_\mu}{dl^2} = \frac{\partial}{\partial x_\mu}\left(\frac{\gamma_{44}}{2}\right)$$

a divergence of some quantity which has to be Newtonian scalar potential of the source, that is, **$G\nabla\varphi$** or **$G\nabla(M/r)$**. This is the source of the **G** factor.

The first term of the equation is the acceleration term and involves the second derivative with respect to time. In relativity **l** is the fourth coordinate, which is **l =ct,** so **$dl^2 = c^2\ dt^2$** and gives the factor **c^2**. So the **G/c^2** factor is fundamental in relativity calculations. This is stated in Einstein's book ***The Meaning of Relativty*** on page 102. Jefimenko has it more modern notation on page 93 of ***Causality Electromagnetic Induction and Gravitation***.

But there is problem with General Relativity. It is a beautiful and perfect theory for gravity in the large. But physics also takes place in the very small, where forces and charges determine the foundations. This is general called the quantum world. General Relativity is simply incompatible with quantum mechanics.

For some fifty years physicist tried to merge the two theories without success. They have given up, since they realized the incompatibility is not a technical mathematical matter, but essential. We have mentioned that the mathematics of General Relativity was real, i.e. did not use complex numbers. Quantum Mechanics is essentially complex, and it is a statistical, not a deterministic theory like General Relativity.

Therefore it is understood that, while General Relativity is accurate in the large for the universe, it is not a complete picture. If the particular structure of its mathematics may require that inertial and gravitational mass be same, it has to be understood that that particular requirement is the product of an incomplete theory.

The problem is that UFOs are NOT in our universe. They carry their own mini negative gravitational universe with them. As far as I know, Einstein never considered negative mass. It is because UFOs are not in our universe, that inertial and gravitational masses are different.

When we see UFOs what we see in effect is an optical illusion. The UFO appears to be in our universe. But because gravitation is not directly visible, we do not see that the gravitation around them is not the gravitation of the universe as we know it. The gravitational fields are radically different. They are actually negative, not positive, as in the rest of the universe. These fields are artificially created by the UFO itself. These fields have different sources and magnitudes.

The magnitude does not arise just from the "charge" connected with the negative mass, the great spin of the negative mass "magnifies" the negative vector potential to 10^{10} times.

Therefore in the equation for Newtonian gravitational force is

$$\mathbf{F_g = \frac{G m_g M}{r^2}}$$

A UFO floats in gravity. It is not attracted by the gravity of earth.
Therefore the force of gravity is zero. we have

$$\mathbf{0 = \frac{G m_g M}{r^2}}$$

and the only way the equation balance is if $\mathbf{m_g = 0}$ since $\mathbf{M}$, the mass of the Earth, is not. The gravitational mass of the UFO appears to be zero because the attractive force on the positive mass and the repulsive force on the negative mass cancel each other out and there is no resulting $\mathbf{F_g}$.. $\mathbf{F_g = 0}$ implies $\mathbf{m_g = 0}$ because of the defining equation.

Likewise the force of inertia is given by

$$\mathbf{F_i = - Gm_i \frac{\partial \underline{A}}{\partial t} = - \frac{Gm_i(a\Phi)}{c^2}}$$

As we have seen in the chapter: "UOFs Have No Inertia", the force of inertia from the universe on the UFO is zero. If there is no inertia then

$$\mathbf{0 = \frac{Gm_i(a\Phi)}{c^2}}$$

and the only way the equation can balance, since $\mathbf{\Phi}$ is finite, is that the inertial mass, $\mathbf{m_i = 0}$. Therefore for a UFO both the gravitational mass, $\mathbf{m_g}$, and the inertial mass, $\mathbf{m_i}$, in our universe are 0. The ONLY reason behind this is the existence of negative mass.

The inertial mass of a UFO cannot actually be zero. If it were, the UFO could only travel at the speed of light and never stand still, which it can.

A UFO has very SMALL inertial mass. As we shall see in the next chapter, it is about ONE MILLIONTH of its normal mass.

INSIDE the UFO, the structure of the UFO IS subject to a scalar gravitational potential. This is the miniscule scalar potential of the negative mass ring. This potential, $\mathbf{-\varphi}$, is both miniscule and negative. But it is finite. Therefore the positive structure of the UFO is immersed in a tiny negative scalar potential.

If body moves against scalar potential, it produces the vector potential, A. No effect.

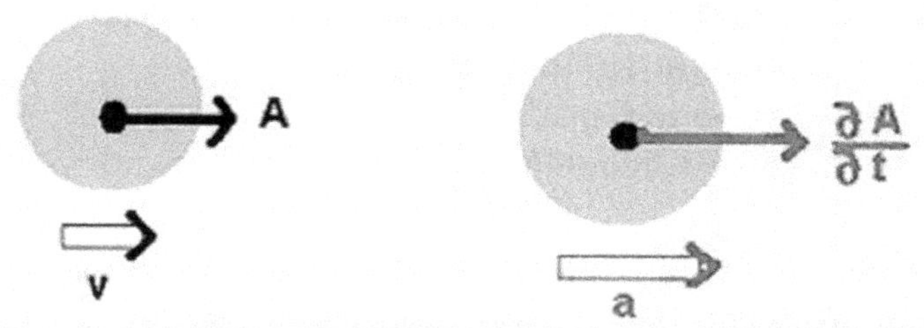

If body ACCELERATES against scalar potential, that creates INERTIA.

$$E = \frac{\partial A}{\partial t} = \text{inertia}$$

And when it moves and is accelerated against this negative field, it will generate a reactive inertial field just as the huge background potential did in space, but it will be much smaller. Calculation shows that it is about one millionth of the field of the background potential.

So the situation is quite complex, and we have to do a detailed analysis of what is going on. Inside the negative gravity bubble of the UFO, the scalar potential of the negative mass in the ring is $\varphi = \mathbf{\Phi/1{,}000{,}000}$, or one millionth of the scalar potential Φ of the universe (see next chapter).

That means the force of inertia inside the bubble $F_i(-)$ is

$$\mathbf{F_i(-) = \frac{F_i(+)}{1{,}000{,}000}}$$

$$\mathbf{F_i(-) = -\ m_i G a \varphi / c^2}$$

$$\mathbf{F_i(-) = -\ \frac{m}{1{,}000{,}000}\ G a \Phi / c^2}$$

as compared to the force of inertia outside in universe, $\mathbf{F_i(+)}$, where the scalar potential is $\mathbf{\Phi}$.

The problem is that we can not SEE the bubble, as gravity is invisible. When we look at the UFO we think, naturally, that it is in our universe. We do not see why the inertial forces felt by the UFO are so ridiculously small. We do not see that they are generated by **φ** and not **Φ**. So that we can describe what we see by our physics, we shift the 1,000,000 denominator from **Φ** to **m**. We say that the inertial mass, $\mathbf{m_i}$, of the UFO is

$$\mathbf{F_i(-) = -(m/1{,}000{,}000)\ Ga\Phi/c^2}$$
$$\mathbf{= -\ m_iGa\Phi/c^2}$$
$$\mathbf{m_i = m/1{,}000{,}000,}\ \text{or one millionth its normal mass.}$$

Therefore the inertial forces on the UFO can be thought of as coming From two sources, the universe and the UFO's negative gravity bubble

$$\mathbf{F_i = F_i\,(+) + F_i(-).}$$

$\mathbf{F_i(+)}$, the inertial force from the universe is zero. It is so because the negative spinning ring of the UFO produces a negative vector potential **A(-)** which blocks the **A(+)** and $\partial\mathbf{A(+)}/\partial\mathbf{t}$ of the universe which normally produce inertia.

We are then left with $\mathbf{F_i = F_i(-)}$, the miniscule inertial force of the bubble, which we ascribe to inertial mass, which then is also miniscule

$$\mathbf{m_i = m/1{,}000{,}000}.$$

Because $\mathbf{F_i(+)}$, inertia from the universe is zero, the UFO can glide through the universe unhindered. It is totally free to accelerate at will and feel no effect from anything. While at the same time the mechanics of its motion are totally governed by the gravitational fields of the bubble with no influence by the powerful gravitational forces of the universe. The UFO moves as if the universe was not even there. And to our eyes the motion of the UFO is such that its inertial mass appears to be about 1 kg.

Therefore to answer Dr. Einstein, the inertial and gravitational masses of the UFO are different purely because of negative mass, something which he never considered.

15 UFO Propulsion

The method that a UFO uses to propel itself is essentially the same as that involved with lifting a car off the ground. In that case the UFO itself moved, and the movement changed the vector potential in the surroundings, and the change in vector potential $\partial \mathbf{A}/\partial \mathbf{t}$ created a gravitational force that lifted the car.

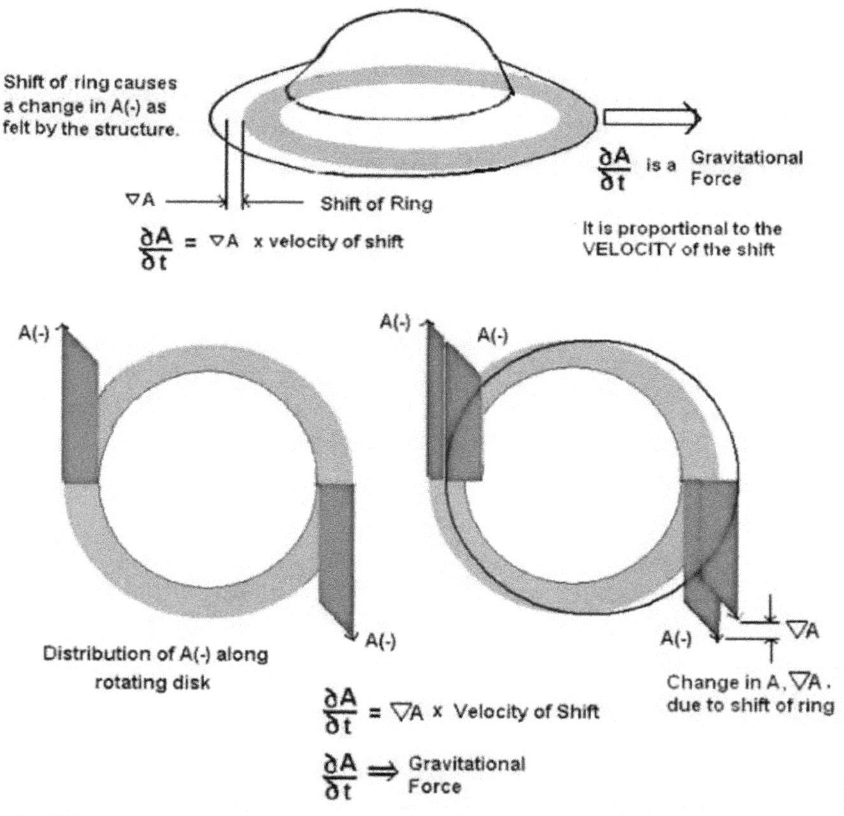

In UFO propulsion, the rotating ring is shifted slightly with respect to the positive mass structure of the craft. That shift creates a $\partial \mathbf{A}/\partial \mathbf{t}$ and hence a gravitational field which the positive mass feels and which then moves the UFO.

We saw earlier that the energy need for a 400 ton UFO to make a 22 acceleration, as was recorded in the radar data in Belgium, would be the equivalent of a small atomic bomb.

That converts to a thrust of

$$\mathbf{F = 400\ tons\ x\ 22g}$$
$$\mathbf{= 4(10)^5\ kg\ \ x\ 220\ m/sec2}$$
$$\mathbf{= 8.8(10)7\ newtons}$$
$$\mathbf{= 88{,}000\ Kilonewtons}$$

or about 16 times the trust of the shuttle. There is no way the spinning ring can generate such force. So how does it happen?

And it turns out that that propulsive force is exceedingly small. How do we know? Very simply. As the UFO accelerates and moves off at great speed, there is absolutely no sign of any change in the UFO. There is no rocket exhaust, there is no sound, the lights on the UFO do not brighten or dim.

By the **Second Law of Thermodynamics**, if the UFO used any appreciable energy to propel itself, the energy would have to be dispersed. It would somehow become visible. Just think of the fiery exhaust of rockets, the flames and roar of fighter jet afterburners, roar of jet engines and turbulence of airliner taking off, and the fiery, smoke and noise of a top fuel dragster heading down the track. The **Second Law** allows for no exceptions.

Then we have the Belgian data, where 22g accelerations by a Big Black Triangle accomplished the acceleration in total silence. What that means is that when the UFO uses energy to propel itself, it is so little, that it is not observably dispersed. How little is that ?

We mentioned the people said the Belgian Big Black Triangle UFOs where about the size of a 747, and that a 747 weighed 200 tons empty. But fully loaded it could be 387 tons. So suppose we take our UFO to be a total of 400 tons. But if you saw the superimposed figures of a 747 and the Big Black Triangle in the chapter on “Inertia”, 800 tons is more likely.
It appears that Big Black Triangles DO disperse energy, and do so constantly and visibly. They have white "lights" with no beams that are constantly on. The “lights” are more likely plasma ports whose purpose is to disperse the heat (energy).

That means the energy use is for something else, not propulsion, as the lights are on even when the UFO is stationary. The energy most likely is to spin the negative mass.

Therefore there appears to be a limit as to how much energy use the UFO can hide. Let us estimated how much this energy is. We will now try to estimate it from observations, i.e. eye witness reports and photographs.

If we look at the Petite-Rochain photograph, the energy in the so called “lights”, looks like each light may be equivalent to 2 or so stadium lights. To get an idea how much energy that is, a stadium lighting typically uses 1,000 and 1,500 watt bulbs. Therefore the Petit-Rochain UFO had the light of 8 stadium lights, or about 10 kilowatts altogether.

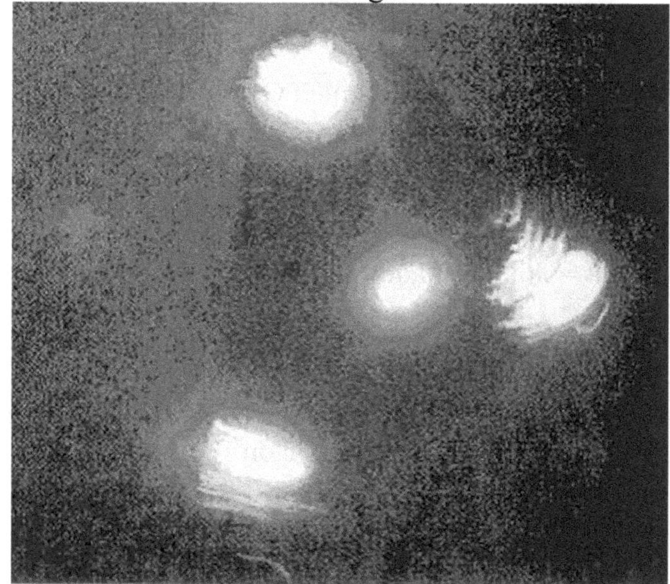

So the continuous use of **10 kilowatts** for spin has to be, and is, dispersed or vented by plasma ports. (See Chapter: “Why Some UFOs Glow”.)

So suppose the UFO could hide the use of this amount of energy, especially if the expenditure lasted only for a short time. A watt is 1 joule per second. UFO accelerations come in short bursts, and the craft then coasts. Bursts are usually five seconds or less. Suppose we could not detect if the an additional **10 kilowatts** was used, but it only lasted 5 seconds. The additional energy release could be absorbed by the body and dissipated slowly and therefore not seen. So that's how we arrive at of **50 kilojoules** of hidden energy use for propulsion.

So if UFO uses as little as **50 kilojoules** for acceleration, how large a mass could be accelerated to **35g**, the largest acceleration from the Belgian data, with that energy?

In the **35g** case, the acceleration lasted 1 second, v = a(t) = a(1) = a, and velocity a(1) = v= 350 m/s (or 786 mph) was achieved.

$$\textbf{Kinetic Energy} = \tfrac{1}{2}\, m_i\, v^2$$
$$\textbf{50 kilojoules} = \tfrac{1}{2}\, m_i\, (350 \text{ m/s})^2$$
$$5(10)^4 = \tfrac{1}{2}\, m_i\, 12.5(10)^4$$
$$0.8 \text{ kg} = m_i$$

So the inertial mass of the 400 ton UFO would have to be about **1 kg**.

More precisely, the apparent mass of **.8 kg** of a **400 ton** UFO would be

$$m_i/m = .8 \text{ kg}/\ 400{,}000 \text{ kg, or}$$
$$m_i/m = 1/500{,}000$$

If UFO had a mass of **800 tons** (which is likely) then

$$m_i/m = 1/1{,}000{,}000$$

Then the ratio of inertial mass inside the bubble to the one in space is approximately one to one million.

How could this be?

We saw earlier that UFOs experience no inertia, that their inertial mass moving in space was "essentially" zero. The negative **A(-)** vector from the UFOs spinning ring blocks the effects of the universal scalar potential $\mathbf{\Phi}$, and the $\partial \mathbf{A}(+)/\partial t$ generated by the acceleration does not reach the craft. Therefore the normal inertia is zero, which we say means the inertial mass is zero for the UFOs motion in space.

So the immersion in the positive universal potential **Φ** has no effect. But the positive mass of the UFO is immersed in another field. And this is the miniscule NEGATIVE potential - φ of the negative mass of the spinning ring.

This immersion in a scalar potential, - φ, generates an inertial force, but because - φ is so much weaker than the universal positive scalar potential **Φ**, the inertial force is very weak. Because we can not see either φ or **Φ**, we think the UFO is in our universe with Φ and ascribe the weakness of the inertial force to a much smaller inertial mass.

We know that the normal inertial field from the universal Φ is

$$\mathbf{E} = \frac{\partial \mathbf{A}}{\partial t} = \mathbf{a}\,\frac{\mathbf{\Phi}}{c^2}$$

We have from Sciama (I) that $\mathbf{\Phi/c^2 = 1/G}$, and since $\mathbf{G = 6.6(10)^{-11}}$ then $\mathbf{E = a\,(1.5)(10)^{10}}$ where **a** is the acceleration.

The formula for the scalar electric potential of ring is given by

$$\varphi_e = 2\pi\sigma(\sqrt{x^2 + R^2} - x)$$

where **R** is the radius of the ring, and **x** is the distance from the ring along a center line and σ is the charge density. The gravitational case would

$$\varphi_g = 2\pi\kappa\sigma_g(\sqrt{x^2 + R^2} - x)$$

where σ_g surface mass density, ρ/**thickness**.

In the negative **A(-)** bubble of the ring, gravitational field E is given by

$$\mathbf{E} = \mathbf{a}\,\varphi_g$$

where φ_g is the scalar potential of the negative mass ring. What we are looking for is the ratio of **E(-)** to **(E+),** that is, how much weaker the gravitational field, and therefore the gravitational inertial force inside the bubble is, compared to outside in the universe.

The ratio would be

$$\frac{\mathbf{E(-)}}{\mathbf{E(+)}} = \frac{\mathbf{a}\,2\pi\kappa\sigma_g(\sqrt{x^2 + R^2} - x)}{\mathbf{a}\,(1.50)(10)^{10}}$$

$$= \frac{2\pi\kappa\sigma_g(\sqrt{x^2 + R^2} - x)}{1.50(10)^{10}}$$

We have with the analogy to electromagnetism that **κ** has the numerical value of 1, so

$$\frac{E(-)}{E(+)} = \frac{2\pi\sigma_g(\sqrt{x^2 + R^2} - x)}{1.50(10)^{10}}$$

If we are looking to find out the relative strength of the potential field very close to the ring, where the positive mass structure of the UFO would be, i.e. the support magnets only inches away, then x is approximately 0, and

$$\frac{E(-)}{E(+)} = \frac{2\pi\sigma_g R}{1.50(10)^{10}}$$

Suppose we assume that any point in the structure sees a potential from contributions of a circle of 1.5 meters in diameter, then σ_g being **1.33 tons/m²**, that is, a 30 meter diameter ring (30 = 2π [D=10 meters]), divided by 40 tons.

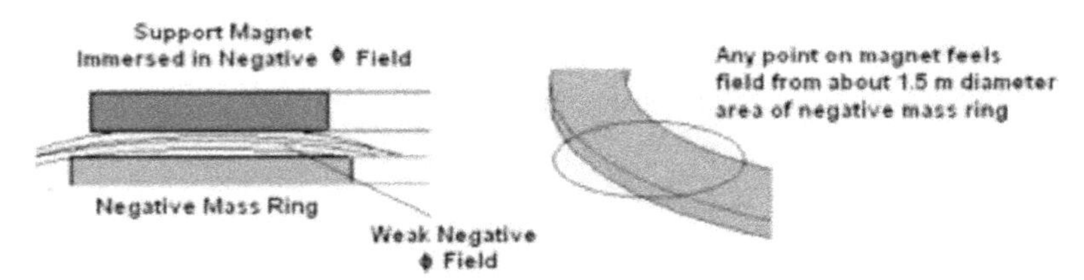

per meter around the ring, the potential would be about

$$\frac{E(-)}{E(+)} = \frac{(6.28)(1.5)(1.33)(10)^3}{1.50(10)^{10}}$$

$$= 1.05(10)^{-6}$$

$$= 1/1{,}000{,}000$$

or a one to a million.

If we now ascribe the weakness of the inertial force not to the field but to the inertial mass, then pretending the potential is the ordinary **Φ**, to compensate, we say the inertial mass is one millionth of the usual mass, that is

$$m_i/m = E(-)/E(+) = 1/\ 1{,}000{,}000.$$

Is it not interesting that we essentially get the IDENTICAL ratio for $\mathbf{m_i/m}$ from the energy use estimate, that the UFO lights do not change during a **35g** acceleration, and from the calculation that the weak inertia inside the bubble comes from the weak negative **φ** scalar field, both approximately <u>one million to one.</u>

This would seem to lend credence to the fact that apparent inertial mass of large UFO is about 1 kilogram.

Now can the shift of the negative mass ring actually propel the craft? Let us calculate.

Suppose the ring is shifted **.1**m or **10 cm**. That would cause a change in the vector potential **∇A** to be **.1A(-)** or **A(-)** reduced by 10%. This is because we assume the material of the ring to be 1 m wide. A 1 meter shift would essentially cause an un-overlap of the magnets and the ring , that is **A(-)** going to zero. So **a .1 m** or **10%** shift would reduce the **A(-)** by 10%.

Suppose the velocity of the shift of the ring to be **1 meter per second**. That would shift the ring .1 m or 10 cm in **.1 second**. Note that the 10 cm shift of the ring in one tenth of a second is a every reasonable figures.

What acceleration would this shift cause? The acceleration would be, with $\nabla \mathbf{A} = .1\ \mathbf{A(-)} = .1(1.5)(10)^{13}$

$$
\begin{aligned}
\mathbf{a} &= \mathbf{G}\ \partial \mathbf{A(-)}/\partial \mathbf{t} \\
&= \mathbf{G\ (3\ rings)\ (ÑA)\ (shift\ velocity)} \\
&= 6.6(10)^{-11}(3)(.1)(1.5)(10)^{13}(1) \\
&= 3(10)^{2} \\
&= 300\ \mathrm{m/sec}^{2} \\
&= 30\mathrm{g}
\end{aligned}
$$

That is, the bone crunching 35g acceleration recorded for the Big Black triangle UFO in Belgium, that would normally have taken a small atomic bomb or 16 times the thrust of the shuttle to achieve, can be achieved by the UFO merely by shifting its negative mass ring(s) in **.1 m** in **just under one tenth of a**

second. And shifting the 40 or 120 ton rings is not a problem, since in the UFO environment they have no inertial mass.

To get an idea what the energies involved in the motion of the UFOs are in more familiar terms are, the **10 kilow**atts coming from white fixed "lights" or plasma ports, presumably to spin the negative mass disks is equivalent to about **13 horsepower**, which is the horsepower of a **200cc** motorcycle. The **50 kilojoule** energy, for (what would normally be **35g** bone crunching acceleration of the huge UFO) is equivalent to a **1 second** power blip from a **60 horsepower VW Beetle motor.**

And now we understand why UFOs move with a zig-zag motion. A movement or acceleration is achieved by displacing the ring in one direction. The UFO then coasts for a while in that direction. To go in another direction another displacement is needed, this now in the new direction. This has the appearance of a zig-zag. The motion of a UFO is therefore a series of independent accelerations in different directions. There does not appear to be a mechanism for making continuous turns.

When you go to the ***UFOCAT*** and investigate UFO turns you find:

1. No entries for "continuous turn"
2. No entries for "smooth turn"
3. 232 entries for "sharp" of "hard" turn
4. 103 entries for "180 degree" turn
5. 381 entries for "90 degree" turn
6. 548 entries for "zig-zag" motion.

This seems to be pretty solid confirmation that the UFO mechanism of propulsion is as we have presented. The displacements of the ring makes the UFO accelerate in one direction, and then another.

16 UFO's Pendulum Motion

The other motion of the UFO we know is the oscillatory or pendulum or "falling leaf" motion. If a UFO engineer were to sit down and design a propulsion system for the UFO, it is difficult to see how the ability to perform "pendulum" motion would a deliberate element of the design. Most likely the pendulum motion is a byproduct of how the system works. As a matter of fact, it may be difficult to design it out. Therefore it casts light on the workings of the entire anti-gravity propulsion system and gives us additional insight.

Let us be clear: although the oscillatory motion of the UFO is described as "pendulum" motion by eye witnesses, it is not the normal pendulum motion caused by gravity. The period of oscillation of a normal pendulum in gravity is independent of mass. It depends only on the length of the pendulum.

$$\mathbf{T = 2\pi\sqrt{L/g}}$$

where **L** is the length of the pendulum and **g** the acceleration due to gravity.

The observed 4 second period of oscillation leads to a length of 6 meters or 18 feet. If the UFO oscillated like a pendulum of length 18 feet its sideways motion would be something like 2 to 3 feet, essentially imperceptible. To appear as a "falling leaf" the excursions of the swing would have to be several UFO diameters. What we now need to analyze how this oscillation can come about in the UFOs anti-gravity drive mechanism.

As we have seen, the negative mass ring is suspended by (most likely) large cryogenic magnets, since we can deduce from the airplane compass disturbance data that UFOs carry such. Addition magnets must act to keep the ring centered. If the UFO is going to shift its furiously spinning ring, it somehow must be able to grab hold of it to move it about. This is probably done again magnetically for the least amount of interference with the spin.

What we have in that case is the following: If the ring is displaced then the magnetic centering mechanism of the UFO now creates

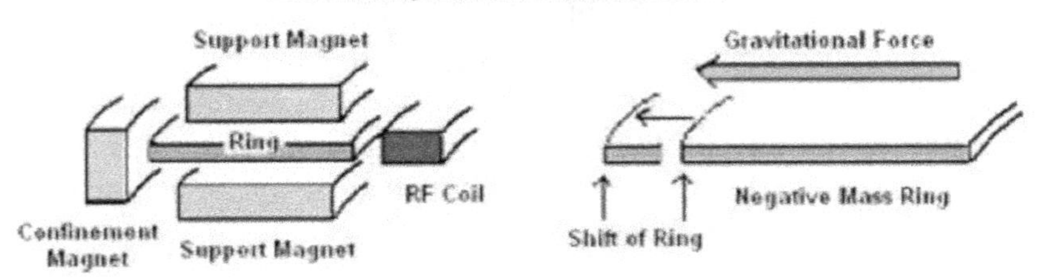

a magnetic force to restore the ring. This is like compressing a spring. An overshoot now creates an opposite compression of the spring. The reverse motion of the ring with respect to of the positive structure now creates an opposite $\partial \mathbf{A}/\partial \mathbf{t}$ and thus a gravitational force which moves the UFO in space in the opposite direction. We have an oscillation.

If we hold the ring fixed, the structure of the UFO must move to relieve the spring. This motion of the structure with respect to the ring creates a change in the vector potential in time and therefore a gravitational force. The gravitational force then moves the UFO in space. When the "spring" is relieved, inertia makes the positive structure of the UFO overshoot, creating an oscillation.

To fit the eye witness metaphor of "pendulum" or "falling leaf" we look carefully at simulations of oscillating weights and try to estimate what a pendulum or falling leaf would look like to an eye witness. It would seem that a period of 4 seconds, or 2 seconds for one half of the swing, and the displacement about 2 to 3 UFO width, or 30 to 50 meters, would look to the eye that the UFO was a pendulum or falling leaf. These parameters could be off by a factor of 2.

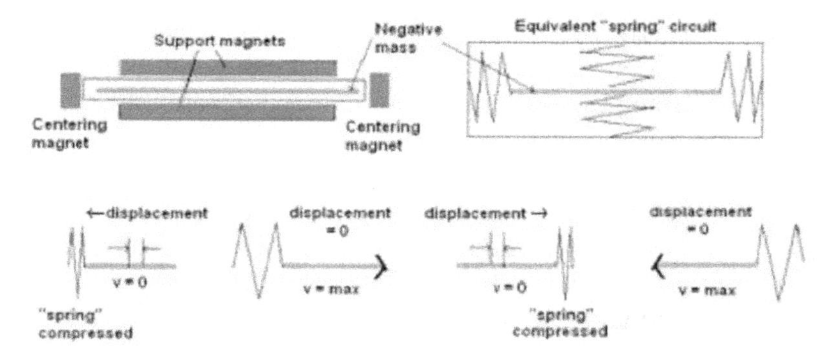

Again, these are just estimates we have tried to extract from the eye witnesses descriptions as "pendulum" or "falling leaf" motion of the seen UFOs.

The anti-gravity propulsion system has two different regimes: The rapid acceleration regime and the oscillatory regime.

In the rapid acceleration regime the ring is displaced quickly in a pulse. Suppose for the **35g** acceleration the ring is displaced **10 cm** in **.1 seconds**. The oscillatory period of the system is **4 seconds**. This is too slow to respond directly to **1/10 second** shift. If we now let the control mechanism return the ring slowly, say **1 second**, then slow reverse movement of the ring, with the shift velocity small, would not induce a gravitational acceleration in the reverse direction. The **4 second** period oscillation would not have a chance to set in. The pulse would have created the acceleration of the craft in one direction and nothing else.

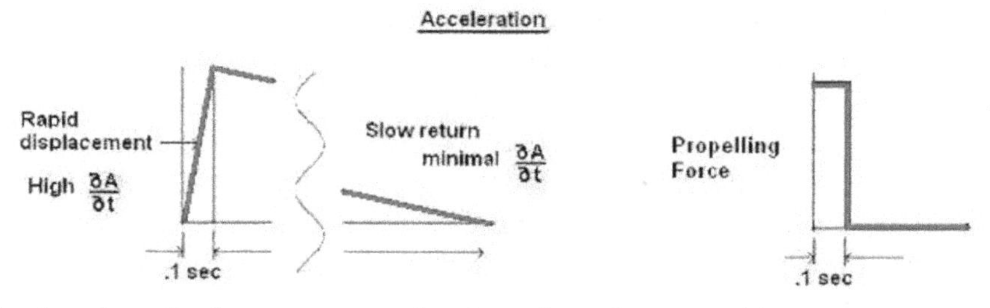

So the ring displacement mechanism therefore would have two modes:

1. Return the ring before oscillation sets in. This is normal propulsion, accelerating the craft in one direction. Returning the ring slowly so it does not cause an opposing acceleration.
2. Not return the ring but let the "spring" "relax" by letting the superstructure move and thus center the ring. This slow relaxation will overshoot and cause an oscillation. In the second case the oscillation could begin and proceed with the 4 second period. In this case what we are doing is giving the oscillatory system an initial condition of a displacement and then letting it is oscillate form there. This could be the Pendulum motion.

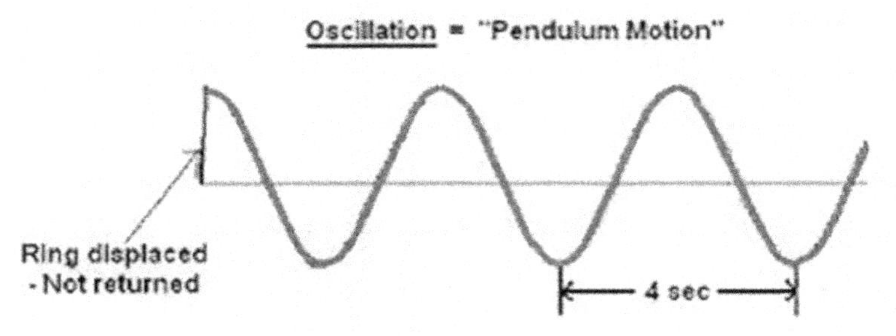

Please note that the gravitational force and spatial acceleration of the UFO is a one way street. The spinning ring creates the vector potential and the force that the positive mass structure of the craft feels, and thus is accelerated. There are no "reactive" forces in the opposite direction.

This miniscule field of the positive mass is merely added to that of the universal background field **Φ** and its effect is blocked by the negative **A(-)** along with that of **A(+)** from **Φ.** It has no effect back on the ring.

The forces on the ring are purely the magnetic restoring forces that connect it to the positive mass. It is these elastic forces together with the finite inertial mass of the structure that make up the oscillatory system of the UFO. Therefore the rapid displacement of the ring, or the oscillations of the ring-structure system, create the spatial motion of the UFO totally "free".

That is, a totally internal motion crates the spectacular accelerations across the sky or oscillations of the UFO. There is absolutely no outward manifestation of these internal motions. The UFO responds externally, in space, to the unseen actions of the UFO's anti-gravity drive.

But what the "pendulum" motion of the UFO does do, is reinforce, or PROVE if you wish, that the inertial mass of the UFO is small, that is, about 1 kg.

The swinging motion must itself involve energy. Suppose we say the swing is about two UFO diameters or 50 meters approximately.

From $\mathbf{f = 2\pi\sqrt{k/M}}$ with **M** the mass and **f** the oscillation frequency we get that an effective spring constant **k** is given by

$$\mathbf{k = (2\pi f)^2 M}$$

The potential energy is then

$$\mathbf{U = \tfrac{1}{2}\, kx^2}$$
$$\mathbf{= \tfrac{1}{2}\, (2\pi f)^2 M x^2}$$

So for a period 4 seconds or frequency of ¼ per second, and an excursion of **25 meters**, we get that the energy involved is

$$\mathbf{U = 360\ M\ joules.}$$

If the mass involved was 40 tons for a small flying saucer, the energy would be

$$\mathbf{U = 360 \times 4(10)^4\ joules \quad or}$$
$$\mathbf{= 1{,}440\ kilojoules.}$$

By the **Second Law of Thermodynamics** such energy would have to be dispersed and highly visible. Since the oscillation is continuous that would mean 1,440 kilowatts of energy expenditure. We know from the above that anything over **10 kilowatts** would be observable. Therefore even the energy of the pendulum motion of a 40 ton UFO would have to be seen.

On the other hand **M** were **1 kg**

$$\mathbf{U = 360\ (1\ kg)\ joules}$$

= 360 joules
= 1/3 kilojoule   would not be visible.

Since no energy dispersal is seen in the "pendulum" motion, this confirms the fact that the inertial mass is small, about **1 kg**.

And we may also have the answer to the puzzle of why **the Dutch Roll** instability sets in. (See the Chapter : "Kinematics of the Upward Falling Leaf Maneuver".) The **Dutch Roll** could be a pilot induced oscillation created by a careless UFO pilot who simply did not center the ring after displacement. An uncentered ring leaves a compressed "spring", which left on its own will cause oscillation.

The missing ingredient in Forward's theory was inertia. Forward was unaware that the inertia in a negative gravitational space is so small. What makes it all possible, what allows the UFO engineers to have the UFO perform all those magical dashing and oscillating tricks, is that with the elimination of inertia the effective inertial mass drops by a factor of a million. The rest is child's play.

17 Summary of UFO Physics

1. UFOs have equal amounts gravitationally positive and negative mass.

2. Because the amounts are equal, the total mass has no "weight". Earth's gravity's push and pull are equal.

3. The negative mass is in form of a ring. It is inside the wide outer lip of a flying saucer.

4. The negative mass is spun at a high rate of speed. This creates an extended range of the negative vector potential **A(-).**

5. The negative vector potential repels the positive vector potential **A(+)** which is derived from the scalar background potential **Φ** of the universe.

6. The time rate of change of **A(+), ∂A(+)/∂t**, is responsible for inertia and **Newton's Second Law**.

7. The negative vector potential **A(-)** creates a bubble from which **∂A(+)/∂t** is excluded. Therefore there is no inertia in the bubble.

8. Because inside the bubble there is no inertia, the negative ring can be spun at speeds near the speed of light because there are no **centrifugal forces** to tear it apart.

9. Because there is no inertia the UFO can make tremendous accelerations with little effort or use of energy.

10. Because in the bubble there is no inertia the **Hall Effect**, acting on electrons, can stop them from flowing and car engines, headlights, and occasionally the lights of towns die out when they are in the expanded bubble of a slow moving flying saucer.

11. The strong vector potential from the spinning ring of a low passing UFO will cause **a ∂A(-)/∂t** on the ground and exert a lifting gravitational force that "levitates" people and objects, sometimes even momentarily lifting cars off the road.

12. The same force that "levitates" is felt by the positive mass of the UFO. If the spinning ring is shifted, the shift creates **∂A(-)/∂t** that is gravitational force that propels the UFO.

13. Flying saucers do not have a mechanism for curved motion. They can accelerate in one direction, and then another. That is why flying saucers have no front or back. They just accelerate in a particular direction.

14. Certain oscillations of the ring produce "**Dutch Roll**", the effect of which is seen as pendulum or "falling leaf" motion of the UFO.

This entire book can be summed up in three sentences:

A) The zig-zag motion of UFOs proves they are not subject to inertia and have (essentially) no inertial mass.

B) Levitation by UFOs means that they are source of powerful negative vector potential **A(-)** which can only be produced by negative mass.

C) Negative mass can be spun at high speed to create the negative vector potential **A(-)** that lifts the car. This **A(-)** also repels the positive vector potential **A(+)** of the universe which causes inertia. That is why UFOs are inertia free.

18 Why the Existence of Negative Gravitational Matter Would Not Be a Big Surprise.

All static fields that we know of arise from charges. Electrons produce the electric field, quarks various color and flavor fields, etc. Therefore the gravitational field is probably produced by some gravitational charge on some subunit of matter. It is estimated that where the quantum mechanical or nitty-gritty of gravity takes place is at the Planck length, some twenty orders of magnitude smaller than any subatomic particles we are used to dealing with.

In physics we have the principles of symmetry and the breaking of symmetry. "Symmetry" used in a very loose sense means that whenever you have a particle of some charge, you are also likely to find a particle of the opposite charge. We have electrons with a negative charge, but electrons with a positive charge exist, and they are called positrons. Protons have a positive charge, but anti-protons have a negative charge. It so happens that the electrons and positrons and the proton and the anti-proton are anti particles of each other. When antiparticles collide, they disappear in an annihilation into photons. Our world is composed of electrons and protons, and not their anti-particles. We don't know why. This so called "symmetry" is broken by some force or principle so that only the particles and not the anti-particles form a stable basis of matter. The anti-particles appear only in decays and accelerator experiments.

In the strong interacting particles we have another kind of symmetry. That symmetry is SU(3) or special unitary (complex) group of three dimensions. If the strong interacting particles had a pure SU(3) symmetry they would all have the same mass. But this symmetry is "broken" by some force so the mass spectrum is spread out. Here the unknown breaking force does not create anti-particles, it just modifies existing ones.

Therefore if gravity comes from some small mass unit we do not yet know, it is very likely that by the analogy with other symmetries that a subunit with an opposite gravitational charge exists, that is negative gravitational matter. This negatively charged gravity subunit does not have to be the anti-particle of the positively charged one. In other words the two could co-exist without annihilating each other.

It is possible, for example, that one in every one hundred gravitational charge subunit is negative. Since for now we can't deal with these subunits individually, the gravitational field we see could be just an average field including the part generated by that negatively gravitationally charged.

There is somewhat of an analogy in nuclear physics. Uranium usually comes as the U(238) isotope, but includes about .72% of the isotope U(235). It is U(235) that is needed to make an atomic bomb, and therefore this isotope has to be separated from U(238) and concentrated. This is very difficult because both isotopes being Uranium have the same chemistry. The only handle one has is the difference in mass, about 1%. Nowadays this is done by centrifuges and about a dozen countries have the ability to do so. Back during World War II they had to use the gaseous diffusion process. If you let uraniumhexaflouride gas pass through porous walls, there will be a tiny difference in the concentration on the other side due to the fact the gas containing the lighter isotope passes through the porous slightly easier. This process has to be repeated over and over again to get the concentration of U(235) and ultimately collected. The collection system is called a cascade. During World War II, the US government built the K-25 plant at Oak Ridge Tennessee. At the

time it was the biggest building in the world, some 125 million cubic feet, 50 acres if five story buildings. Maybe something like that has to be done to separate out negatively charged gravitational mass.

And one scientist, F. Winterberg, has theorized that quarks actually contain negative matter. (1)

The other possibility is that the gravitational subunits could be manipulated artificially and the gravitational sign of the charge just flipped. We can now in the lab make transuramic elements that do not exist in nature. They have a very short half life, but it can be done. We can do tricks like the following: we can make positrons and anti-protons and combine them into

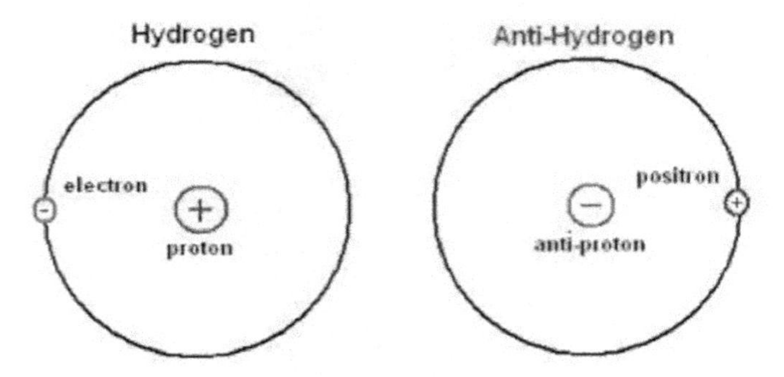

antihdyrogen where the normal charges of the electron and proton are reversed. Of course antihydrogen doesn't last very long, as it annihilates when it hits the walls of the vessel. So the UFO people may have found a way to flip the sign of gravitational charge.

All of the above are speculations and arguments by analogy. But on their basis it would not be wildly out of the ordinary in physics for negative gravitational matter to exist.

1) F. Winterberg, “Quarks, magnetic monopoles, and negative mass.”, Lett. Nuovo Cimento, Vol. 13, Ser. 2, 1975, pp.697-703.

19 ARTICLE: The Fundamental Equation of a Flying Saucer, With Applications.

ABSTRACT

We derive a simple fundamental equation for a flying saucer. The equation is:

$$\mathbf{v/G = \kappa Mfa^2/r^2}$$

where the **v** is the velocity of the craft, **G** the gravitational constant, **M** the negative gravitational mass , **a** the diameter of the rotating ring, **f** frequency of rotation **r** the radius of the negative vector potential bubble, and **κ** a constant of value 1 with units of m/kg. From the equation it is possible to deduce that: 1) inertia can be eliminated, 2) the minimal size of possible for a saucer, in agreement with eye witnesses reports, 3) explain the EM effect of car headlights and engines dying near UFOs, 4) explain why UFOs do not cause sonic booms at supersonic speeds, 5) and explain reports of levitation and loss of car control near a UFO. Our only non-standard assumption is that the saucer posses negative mass, an assumption which is NOT contrary to known physics principles.

Eliminating Inertia

Bondi (1) has show that the existence of negative gravitational mass does not violate general relativity. This conclusion has not been refuted.

Sciama in 1957 did the first calculation that showed that the total matter in the universe caused inertia (2). His calculation was non-relativistic. Einstein, although he never explicitly included Mach's Principle as part general relativity, says on page 102 of ***The Meaning of Relativity*** that inertia was mediated by the time derivative of the vector potential **A**, that is, by gravity (3). In a recent paper entitled "On Mach's principle: Inertia as gravitation", a Spanish group did a fully relativistic analysis with linearized Einstein equations. They used the time derivative of the vector potential, $h_{0i,0}$, but also a small scalar counter term $h_{00,i}$, retarded potentials, and included consideration of dark energy. They came within 10% of verifying **Mach's Principle**. Therefore the principle that gravitation causes inertia is well established.

Following Sciama (4) and Jefimenko (5,6) we have the vector potential

$$\mathbf{A} = \mathbf{G/c^2 \int \rho v/r \, dV}$$

where ρ is the mass density and v its velocity. To find the vector potential created by the remote stars postulated by **Mach's Principle**, Sciama carries out the integration over the mass distribution of the entire universe, assuming the universe to be receding with velocity **-v**. The **v** being constant comes out of the integral

$$\mathbf{A} = \mathbf{vG/c^2(\int \rho/r \, dV).}$$

With

$$\mathbf{\Phi} = \mathbf{G(\int \rho/r \, dV)}$$

We are left with

$$\mathbf{A} = \mathbf{v\Phi/c^2}$$

where **Φ** is the universal scalar gravitational potential. Sciama then changes coordinates and has the test particle move with v while the universe is stationary. The recent calculation (4) also uses this technique. We are going to interpret this equation to mean that the particle id subject to the **A** vector as it moves through the universes scalar potential.

Positive and negative masses repel each other through their fields, and the field has to be zero on the boundary. If a region of space existed around a gravitationally negative mass, the scalar potential would be negative and the **A** vector derived from this field would be anti-parallel to the velocity

$$\mathbf{A} = \mathbf{- |\varphi| v/c^2}$$

as opposed to parallel, $\mathbf{A} = \mathbf{\Phi v/c^2}$.

This is to be expected as **Φ** and **A** form a four-vector in relativity

$$\mathbf{A} = \mathbf{(A,\varphi)}$$

so if **φ** is negative, so is **A.**

These parallel conditions are mutually exclusive; you cannot have **A** parallel and anti-parallel to **v** at the same time. Therefore the positive and negative gravity spaces would have to be disjoint.

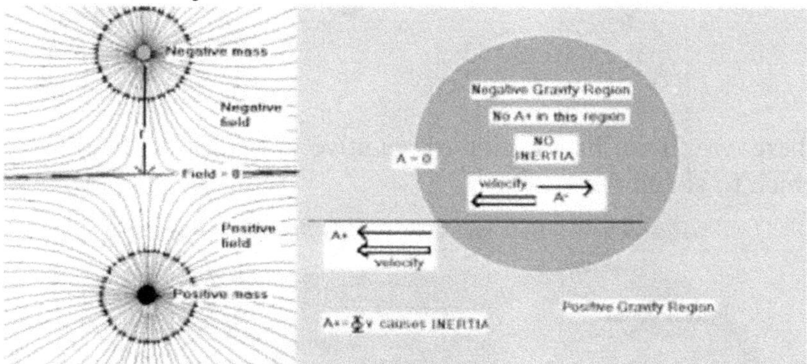

A vector potential would tend toward zero as we approached the boundary of the negative gravitational space, become zero at the boundary, and then become negative (or anti-parallel) inside the negative gravitational space. The parallel **A(+)** vector could not exist in the negative gravitational space. Therefore the **A(+)** vector could not mediate inertia in the negative region.

This then provides the possibility of shielding from inertia. The possibility Of shielding from inertia is derived from the corollary to **Mach's Principle**. If the "distant stars", that is, the universal scalar background potential, **Φ**, causes inertia, then the corollary is that in the in the absence of the effect of the "distant stars", or the effect of **Φ**, there is no inertia.

We see that as the disparity between the size of positive and negative masses increases, the field around the negative mass shrinks.

The problem is that any finite amount of negative mass, tons or even hundreds of tons, would have a gravity field that would be miniscule compared to the field of the astronomical quantities of gravitationally positive mass. As the positive mass increases the negative field would be swamped and actually be confined to within the negative matter.

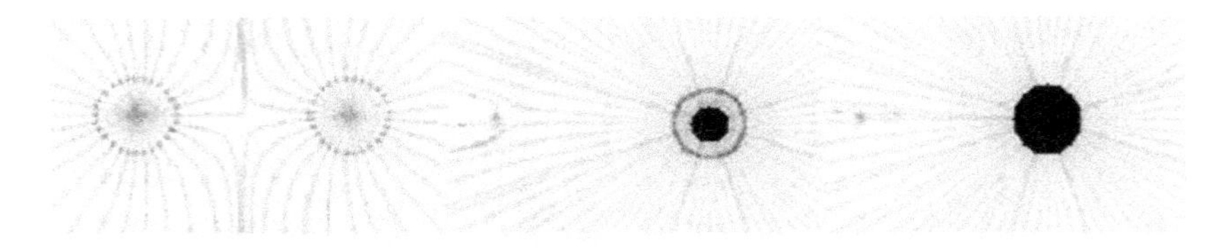

There would not be a region of negative gravitational field in which to shield the UFO.

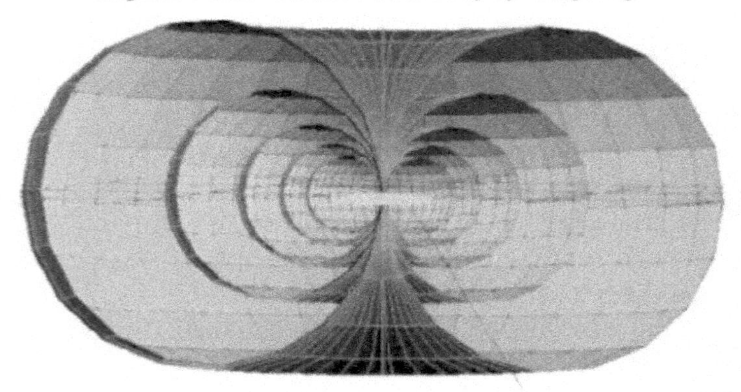

.

If we take a negative ring and spin it, we create a magnetic dipole If we look at the definition of **A**, we see that there is a possible multiplier, the velocity **v**.

$$\mathbf{A} = \int \rho\, \underline{\mathbf{v}}/\mathbf{r}\, \mathbf{dV}.$$

A field. If we spin the ring faster and faster, over a million revolutions per second, we can get a finite field. Therefore we create a bubble in space of

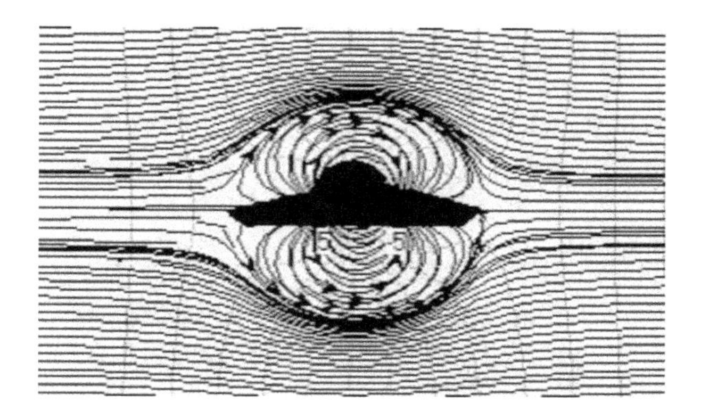

negative gravitational fields. If an object were to be within this bubble, it would be shielded from the positive vector potential arising out of the universe's background scalar potential. In other words, the object would be shielded from the fields that create inertia; it would not be subject to inertia.

When we spin the mass, the mass is now within the bubble, and therefore has no inertia. Without inertia the inertial mass and moment of inertia of the spinning body are zero, and consequently there are no centrifugal forces to tear it apart. The possible rates of spin can be very high, the only limitation being that the rim velocity of the spinning body could not exceed the speed of light.

<u>Deriving The Equation</u>

We now use the fact that a spinning ring of negative mass can create a negative gravitational field to derive the fundamental equation for flying saucer.

At the boundary between the positive and negative gravitational spaces the total vector potential is zero. Labeling the **A** vectors with + and - to keep track of where they operate

$$\mathbf{(A+) + (A-) =}$$

$$\mathbf{(A+) = -(A-)}$$

On the left hand side of the equation we have the vector potential of the universe which is given by

$$\mathbf{A(+) = \Phi v/c^2}$$

where v the velocity of the craft.

The flying saucer generates a negative vector potential **A(-)** by spinning ring of negative mass. The spinning ring is a gravitational dipole that generates the **A(-)**

$$- A(-) = \ /r^2$$

Since the equations of the gravitational weak field approximation are identical to Maxwell's, we can take over this equation. The normal magnetic dipole of a current loop is given by, where I is the current.

$$= \ _0 I(\text{Area})$$

The shape of the negative mass in the saucer is probably a flat ring, something like a frisbee. We will simplify our calculation by replacing the flat ring with a thin massive ring of linear mass **λ**. We replace $_0$ by **κ,** which we will discuss in a NOTE below, and I by a current of mass, $I_m = \lambda v$.

$$\lambda = \text{mass/circumference}$$
$$= M/(2\pi a).$$

The area is πa^2. Therefore the mass current is

$$I_m = Mv/(2\pi a)$$

So gravitational magnetic dipole moment will be

$$\mu = \kappa\, Mv/(2\pi a)\, \pi a^2$$
$$= \kappa\, (M/2) v\, a$$
$$= \kappa M v a/2$$

The result on the right hand side for the negative vector potential is

$$- A(-) = \ /r^2$$

Because we have a negative mass (the sign of M is negative) which cancels the minus sign in front of A(-) and therefore

$$A(-) = \kappa M v\, a/(2r^2)$$

We can express the formula in terms of the rotation frequency f. The velocity $v = a\,\omega = a\,(2\pi f) = 2\pi a f$, where ω is the angular velocity. Then

$$A(-) = \kappa M(2\pi a f)a/(2r^2)$$
$$= \kappa\pi M f a^2/(r^2)$$

The left hand side of the equation **A** of the universe is

$$\mathbf{A(+) = v\Phi/c^2 = v/G}$$

since from Sciama (4) $\mathbf{\Phi/c^2 = 1/G}$

We have the derived the fundamental equation for a flying saucer:

$$\mathbf{v/G = \kappa\pi Mfa^2/(r^2).}$$

The equations for the vector potential **A** are vector equations but for our formula we need consider only the scalar part as the vector aspect is only a spatial component of order unity.

APPLICATIONS

1. UFO Without Inertia

In order to test whether a flying saucer is able to overcome inertia, we apply the formula to the worst case scenario, the smallest flying saucer reported with a ring radius of say 5 meters, or total diameter of 30 feet. Because the generation of the negative vector potential A(-) depends on the negative mass which has to be rotated, a larger UFO which could have a more massive ring, could possibly not have to rotate the ring quite as fast and still be able to overcome inertia. Also the speeds recorded by radar in Belgium in 1990 where for Big Black Triangle type UFOs.

We are limited by size (and mass) of the ring and the rotation frequency because the rim velocity cannot exceed the speed of light. The question now is, what kind of a rotational speed do we need for the spinning ring to create a field that equals the field from the universal, scalar potential, repelling it, and create a space in which the UFO can hide

In any case, if a minimal flying saucer, 30 feet in diameter, the smallest reported, is able to overcome inertia, then we know all other, larger UFOs can also because the criteria are less stringent for them.

The formula for the vector potential

A(-) = /r²

is called the dipole approximation. It only works for distances r large compared with dipole radius **a**. It is clear that the formula has a singularity at **r = 0**. That is, the formula assumes that the dipole is a point. This is not so. The dipole is actually a finite loop.

The exact formula for **A** from a current carrying loop involves elliptic integrals. Examining the exact formula near **r = a**, the elliptic integrals turn out to be approximately unity. Also a power series expansion for **A(-)** near the ring shows that the vector potential is very close in value to dipole moment **.** We are therefore going to make the approximation, that nothing drastic happens near the value **r = 1**. We will therefore take r to be 1 when we are near the dipole, and essentially at the dipole the vector **A(-)** and the dipole moment have the same numerical value, though of course the dimensions are different.

On the left hand side of the equation we see that it is proportional to the velocity of the craft. From Belgian radar we know speeds of over 1,000 miles per hour have been recorded. If we take v = 1 km/sec or 2,200 mph, that should cover all UFO speeds. So the left side would be **A = 1,000 m/sec/G,** *or* **A= 1000/G**. (7)

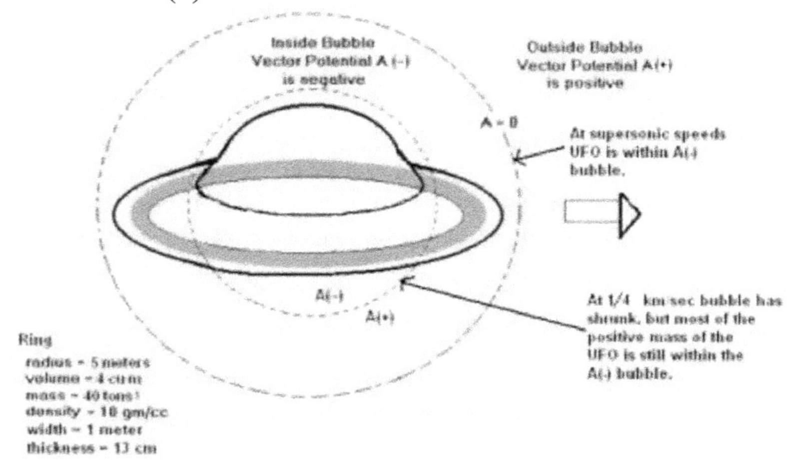

Therefore the boundary between the positive potential **A(+)** and the negative potential **A(-)** will be approximately at r = 1 meter when the saucer is traveling 2,200 miles per hour.

We find then that with,

$$\mathbf{v/G = \pi M f a^2}$$

the equation balances. On the left

$$\mathbf{v/G = 1000/6.6(10)^{-11}}$$
$$\mathbf{= (10)31.5(10)^{10}}$$
$$\mathbf{= 1.5(10)^{13}}$$

And on the right

$$\mathbf{\pi M f a^2 = (3.14)4(10)^4 5(10)^6 (5)^2}$$
$$\mathbf{= 1.5(10)^{13}}$$

Therefore the saucer is within a negative potential bubble where the positive potential **A(+)** can not reach. The bubble is velocity dependent and shrinks with increased velocity. The saucer is in the bubble even at 2,200 miles per hour. This latter statement is an approximation, and is not quite true. The question is discussed in the Chapter on "The Shape of a Flying Saucer".

2) Minimal Size of Saucer

If try to use some other geometry for the negative mass, a disk for example, the vector potential A(-) is not strong enough to overcome inertia. The calculations are shown in the chapter on the "Gravitational Engine."

It also turn out that for this 30 foot diameter saucer to produce enough vector potential A(-) it has to rotate at one half the speed of light. That is now pushing the boundaries of the possible, since for that speed the relativistic mass increase of the negative mass is 15%.

Therefore the 30 foot saucer is the minimal flying saucer possible based on limitations imposed by physics. Interestingly enough, the 30 foot flying saucer has been the smallest saucer reported in eye witness reports.

3) EM Effects

By our formula, when the UFO slows down, **r** increases. Therefore the inertia free region around the UFO expands. It is this inertia free region that cause EM effects, which we will discuss shortly. How far away from the saucer have EM effects been reported? There are 5 reports in the ***UFOCAT*** at .4 miles, one at .78 miles, and 2 at 1.53 miles. There are two more at a greater distance, but they are from airplanes. Unfortunately most the 1,820 EM effect reports do not give distance, and even those given may not be accurate. As a conservative estimate let us say radius of the inertia free zone **.78 miles** or **1, 250 meters**.

Theoretically when the UFO is stationary and **v = 0**, there is no **A(+)** generated, and value of the for the left hand side of the equation is zero. It is possible that there is a residual or background vector potential A(+) even at zero velocity. Let us use the EM radius data to estimate such a possible potential.

Since

$$v/G = 1.5(10)^{13}/r^2$$

$$v \times 1.5(10)^{10} = 1.5(10)^{13}/r^2$$

$$v = 1{,}000/r^2$$

.

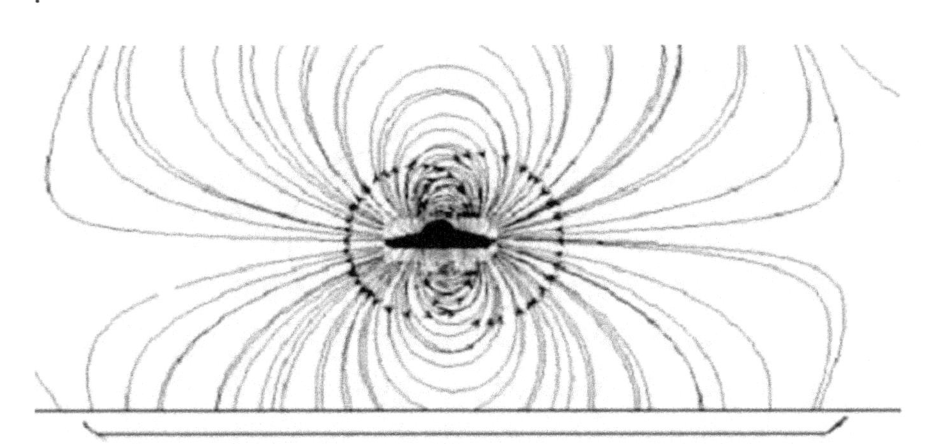

1.5 miles of near zero inertia even for a small flying saucer

Converting that that distance to an equivalent velocity,

$$\mathbf{v = 1{,}000/(1250)^2}$$
$$\mathbf{= .00064\ meters/\ second}$$

We see that the UFO is essentially stationary

Then the possible residual or background **A(+)** would have to be less than

$$\mathbf{A(+) = \ v/G = 6.4(10^{)-4}1.5(10)^{10}}$$
$$\mathbf{= 9.8(10)^6\ kg\ sec/meter^2}$$

If the saucer is at low altitude the bubble breaks the ground plane and witnesses find themselves in an inertia free zone. The movement of limbs have slow acceleration. Our muscles are designed to resist the pull of gravity. We normally do not feel inertial effects. If they were missing, we would not notice.

The story with sub-atomic particles is the reverse. Gravitation on their mass is negligible. Electrons in metals behave as a free gas. Their high momentum and kinetic energy are critically dependent on inertial mass. If that changed, there would be noticeable effect. And that is what happens in the UFO's inertial free zone.

A window alarm attached to the window frame has a small magnet near it attached to the sash. If the sash is opened, the magnet is moved away and the alarm sounds. This is because of the **Hall Effect**. The window alarm uses a

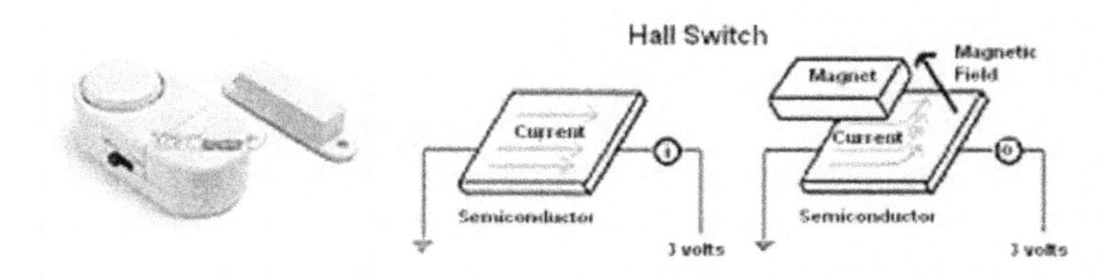

semiconductor where the structure of the semiconductor reduces the effective mass of the electrons to a few percent of their normal value. Since the momentum of the electron is,

m_iv, inertial mass times the velocity, its momentum is now small. The charge of electron remains unchanged, so a magnetic field can interact with the charge to easily bend the path of the electron with weakened momentum. The small magnet can deflect the current by the **Hall Effect**. The magnet bends the path of the electrons in the semiconductor. The electrons are diverted. They do not reach the anode. Remove the magnet and the current flows and the alarm sounds.

The mysterious EM effect of car engines and headlights dying near flying saucers (10) now has a simple explanation. In a semiconductor the electron waves interfere with reflections off the lattice. The effect is that inertial mass of electrons appears to decrease. In other words if you want to write a theory of electron behavior in a semiconductor, what you do is assign an effective mass, **m_e** , to the electron that is less than its actual mass and the theory works. (See for instance Kittel ***Introduction to Solid State Physics*** (11)). It turns out that a correct theory just posits electrons to have an effective **mass** = **m_e** as low as 1 to 2% of the value when free.

Crystal	Electron m_e/m
InSb	0.015
InAs	0.026
InP	0.073
GaSb	0.047
GaAs	0.066

An electric field will move electrons. If you now add a magnetic field The electrons will move in a circle. The diameter of the circle will be determined by the relative strength of the fields and the charge to mass ratio of the electron. The charge to mass ratio is **e/m.** If **m** is small, then the charge to mass ratio will be large. Since the magnetic field exerts a force on the charge to bend the motion, and the momentum or inertial mass of the electron wants to make it go straight, a high **e/m** means that electrons path will be more easily bent. Since the momentum of the electron is, **m_iv,** inertial mass times the velocity, its momentum is now small. Since **e/m$_i$** ratio can be 50 to 100 times larger in the semiconductor than in a conductor, semiconductors can be used to bend the path of electrons even with small magnets. This is called the **Hall Effect**.

When a car finds itself within the UFOs anti-gravity bubble, the inertial mass of electrons in the wiring is essentially reduced to zero since they can no longer feel the effects of the positive vector potential of the universe, **Φ**. Therefore their charge to mass ratio increases. The electrons behave as if they were in a semiconductor, but even more so, since their inertial mass is closer to zero. Therefore any stray magnetic field will divert their motion, and stop the current from flowing.

And such stray fields exist everywhere. For example in typical 14 gauge wiring used in automobiles, a 10 amp current will generate a magnetic field

Magnetic field
B = 25 Gauss
in wire carrying
current I = 10 Amps

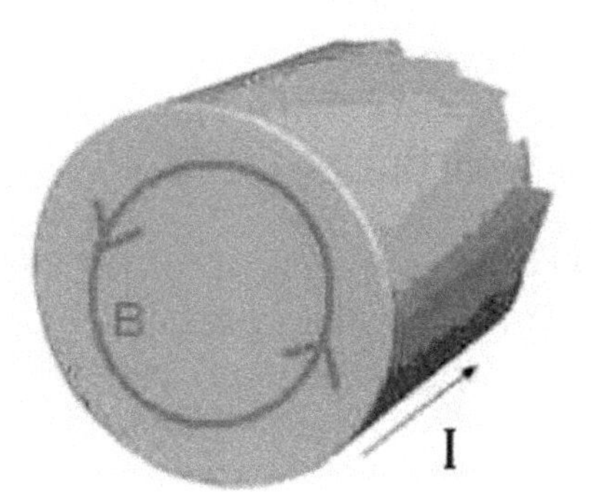

inside the wire of about 25 gauss, the same as strength as that of little magnet that comes with the window alarm. Therefore the flowing of a current itself produces a magnetic field, and this field could shut down the current.

Therefore the EM effects of flying saucer are caused by the **Hall Effect**, which diverts the electrons, prevents current from flowing, and kills car headlights and engines. When the flying saucer leaves, the negative field is removed, and the devices spontaneously function again. (There are 475 entries in the ***UFOCAT*** for dying headlights and engines)

There are also reports of witnesses experiencing spontaneous paralysis under these circumstances (10). Nerves work by transmitting electrical signals mediated by electrons, much as current in a wire. The nerve axons in the limbs are 3 to 4 feet long. This means that the **Hall Effect** has an opportunity to interfere with transmissions of nerve signals at many points along the axon. A closer analysis, I believe, will show that the **Hall Effect** is also responsible for the reported paralysis. (There are 109 entries in ***UFOCAT*** for paralysis.)

It is interesting to note that airplane engines near flying saucer do not die but only sputter. The reason may be is that airplanes engines use magnetos to generate the spark. Magnetos are electro-mechanical devices that produce high voltages connected to spark plugs with short wires. The high voltage and short wiring path may prevent the **Hall Effect** from shutting them down.

4) No Sonic Boom

In 1990, Belgian radar recorded UFOs traveling at speeds in excess of 1,000 miles per hour. The official Belgian Air Force report noted that no sonic booms were heard on the ground. The explanation is now simple. The air molecules around the UFO are within the inertia-free bubble created by craft. Because the air molecules effectively have no inertial mass, they do not resist being pushed out of the way. They do not pile up in front of the craft. It is this pile up which causes the shock wave that ordinarily generates the sonic booms.

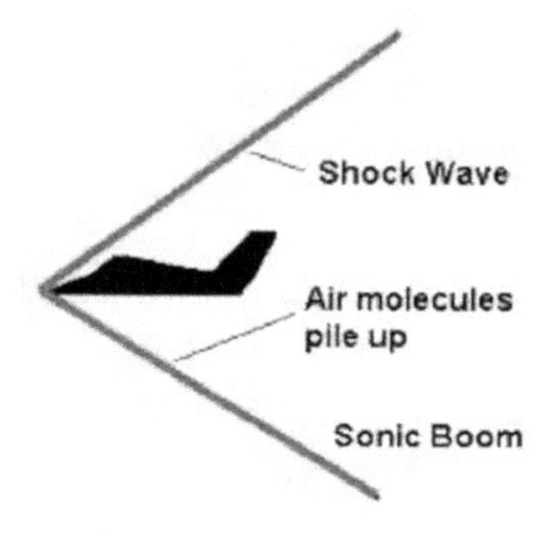

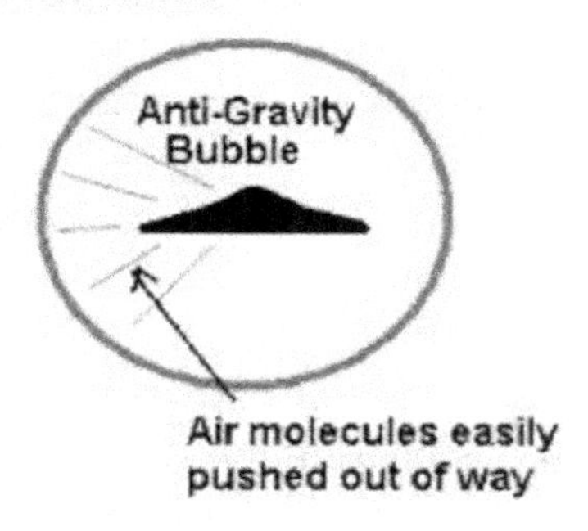

This also explains why flying saucers do not have aerodynamic shapes, yet they do not have problems with air resistance, even traveling at over 1,000 miles per hour. The air molecules, being in the inertia free bubble around the craft simply do not resist being easily pushed out of the way.

5) <u>Levitation and Loss of Car Control and Lifting Car off the Road</u>

There are occasional reports of levitation by eye-witnesses (10). The loss of car driving control often accompanies such reports. What that means is that some of the weight is taken off the tires so that they do not provide sufficient traction to steer the car. (There are 300 entries in ***UFOCAT*** for levitation.) The levitation does not always mean that the witnesses are actually lifted off the ground but that they experience lifting force in that direction.

These are not inertial events as they are not the result of movement but are forces experienced with the subject at rest.

We have here something entirely new: the saucer is the source of an independent gravitational force. A finite object, the negative mass ring, by being spun creates a vector potential on par with the vector potential that comes from the scalar potential of the universe. We are not dealing here with Newtonian gravitational attraction. The gravitational force here is a result of motion, and is entirely derived from the vector potential. These fields that have the palpable power, to levitate people and objects, to interference with the steering of vehicles, and occasionally lifting the entire vehicle off the ground.

The forces appear to be of short duration and only when a UFO is passing. Only two distances to the saucer are listed in ***UFOCAT*** for these encounters, one at 16 and the other 64 feet.

The negative vector potential field of the saucer has a $\mathbf{1/r^2}$ gradient. If a flying saucer were passing, the vector potential field would be changing with time because of the gradient. This means that there is a time derivative of the vector potential which by the analog of Faraday's Law produces

$$\mathbf{F} = \mathbf{m}\, \partial\mathbf{A}/\partial t$$

a gravitational force.

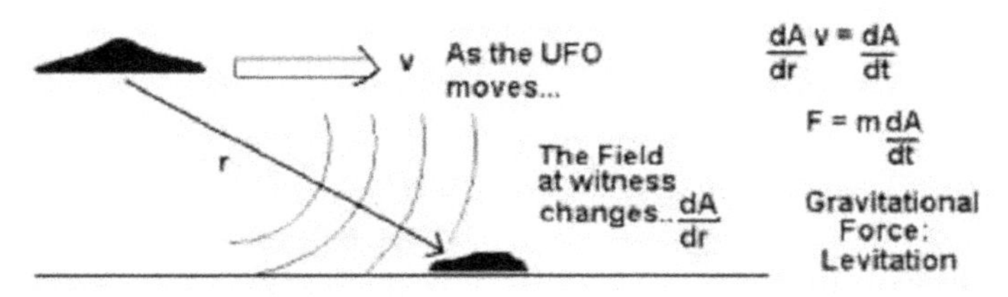

We note that because the saucer is moving the bubble shrinks and these gravitational effects can be felt only near the saucer.

We have $\mathbf{A} = \mathbf{v}/\mathbf{G}$

$$\frac{\partial \mathbf{A}}{\partial \mathbf{t}} = \mathbf{a}/\mathbf{G}$$

By multiplying through by **G**, we get an acceleration

$$\mathbf{a} = \mathbf{G}\,\frac{\partial \mathbf{A}}{\partial \mathbf{t}}$$

which for gravity would be 10m/sec/sec on the earth’s surface.

As the flying saucer moves, because of the gradient of the **A** field, there would be change of the A vector with time:

$$\frac{\partial A}{\partial t} = \frac{\partial A}{\partial r}\frac{\partial r}{\partial t}$$

$$= \frac{\partial A}{\partial r}(rv)$$

where rv is the relative velocity, which we take to be, say, 20 m/sec. The vector potential **A** is the dipole moment divided by r^2

$$\mathbf{A} = \ /r^2$$

Taking the very stringent case of a small suacer or 30 foot diameter, where the dipole moment would be $1.5(10)^{13}$, then gradient of A becomes

$$\frac{\partial A}{\partial r} = -2 \ /r^3$$

$$= -\ 2x\ 1.5(10)^{13}/r^3$$

And taking the relative velocity **rv = - 20**

$$\frac{\partial A}{\partial t} = \frac{\partial A}{\partial r}(rv) = 3(20)(10)^{13}/r^3$$

$$= 6(10)^{14}/r^3$$

The acceleration produced by this field at a distance of **r** from the UFO is

$$a = G\frac{\partial A}{\partial t} = 6.6(10)\text{-}^{11} \text{ x } 6(10)^{14}/r^3$$

$$= 4(10)^4/r^3$$

At **20 meters**, or about **60** feet, the lifting would be

= 40,000/8000
= 5 meters/sec/sec
= 50% of body weight lift.

At **7 meters**, or about **20 feet**, the lifting would be

= 40,000/343
= 12 meters/sec/sec
= 120% of body weight lift

enough to lift an object off the ground.

A car would be lifted off the ground, because the upward lifting force, 120% of body weight, would be greater than its weight or downward gravity pull at 100% of body weight.

So if witnesses have a flying saucer pass by at a range of 100 feet or less, even at a very reasonable speed of 20 m/sec or 44 mph, they can expect to experience gravitational forces lifting ("levitating") them, of the steering of their car, where the saucer's spinning ring created the force. But if the saucer passed very close, one can expect the car to be lifted off the ground. Because the lifting force is sometimes at an angle, reports have the front or back only of the car lifted, or the car lifted on two wheels. But sometimes it is lifted entirely in the air.

UFO Design

Just a few words on what the equation tells us about UFO design. The familiar wide flange at the base of flying saucers may be there because the field generated depends upon the mass being moved at close to the speed of light. So the ring has to be made as large as possible, and the flange is there to cover the wide rotating disk.

Just as the circular shape of the flying saucer suggested a single rotating ring, so the shape of Big Black Triangle suggests three rings

The Big Black Triangle UFOs typically measure hundreds of feet. They have multiple rotating rings. Having several dipoles in parallel results in a more oblate combined field, more closely wrapped around the craft. The bubble from this field, being more flat than the spherical field of the single dipole rotor, would be less likely to break the ground plane even at low altitudes.

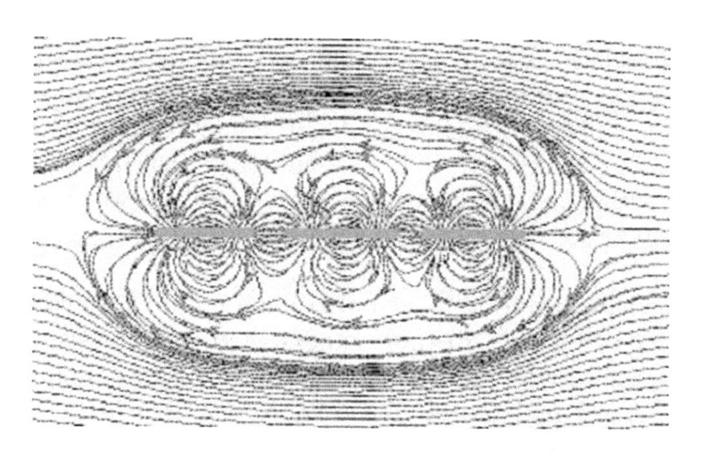

Indeed, there are few reports of side effects for such UFOs. Of the 7,000 Big Black Triangle sightings in ***Night Siege*** there were only two EM effects (12). In one, the UFO was skimming by the top of a bridge tower when the lights on that tower went out. This indicates that one has to very near a Big Black Triangle UFO to break into its anti-gravity bubble and experience an

NOTE on constant factor Kappa

The definition for the gravitational vector potential **A** in texts is

$$\mathbf{A = G/c^2\ \rho v/r\ dV}$$

Einstein in the ***Meaning of Relativi***ty on page quotes this.

The coefficient $\mathbf{G/c^2}$ is the inverse of a huge number, $\mathbf{10^{-29}}$.
Perhaps we can understand this, because Sciama(4) has

$$\mathbf{\Phi G = c^2}$$

or $\mathbf{A = 1/\Phi \int \rho v/r\ dV}$.

In effect, the calculation for vector potential has to be

"normalized"by the background scalar potential **Φ**. This is an expression of **Mach's Principle** that "the distant stars" have a profound effect locally. The consequence is, that to get any measurable result for **A**, one needs to integrate over the entire mass of the universe, which is what Sciama (4) and Martin (2) have done. The **A** from the integration of the universe is strong enough to have the measurable effect, namely inertia.

But what it also means is that gravitational fields from finite or even astronomical objects like planets, relying on this "normalization" will ALWAYS be infinitesimal. This is in direct conflict with eye witness reports that flying saucers generate considerable gravitational fields, to the point that sometimes they can lift a vehicle off the road.

We have a stark choice: either we go with "normalization", in which case ALL the eye witness accounts must be false, or we must discover why the "normalization" is not applicable in this case, and find another value for Kappa, that makes the results of the calculation of gravitational **A(-)** finite.

What we need to realize is that negative gravitational region created by the spinning negative gravitational ring is a "mini" universe, which is different from our universe.

In our universe all mass is gravitationally positive. Each body has it's own positive scalar potential which adds cumulatively to that of all others, creating in sum the universal background scalar potential **Φ**. This background potential pervades every nook and cranny of the universe. It is also responsible for inertia, **as Mach's Principle** and the work of Sciama and Martin et.al. has fairly conclusively shown.

But the negative gravitational potential **A(-)** our spinning ring produces repels the positive vector potential **A(+)** that arises from **Φ**. It does not block **Φ** itself, but **A(+)** derived from Φ. This negative region is only meters wide. There are no other fields in this region. And because **A(+)** is blocked, **Φ** has no influence. It is because **Φ** is not relevant in this region that we do not "normalize" by it.

This situation in the negative mass region is very similar to the case electromagnetism. The positive and negative electric charges of protons and electrons in matter cancel each other out. Matter does not create a cumulative electric potential. When we calculate the magnetic vector potential of a current loop, we need to consider no other field, we do not "normalize" with anything. The formula for **A** is merely

$$\mathbf{A_{em} = I(area)/r^2}$$

In our calculation of **A**

$$\mathbf{A = G/c^2 \int \rho v/r\, dV}$$

we replace $\mathbf{G/c^2}$ or $\mathbf{1/\Phi}$ by **Kappa.** What value does **Kappa** have? By our analogy with electromagnetism above, **Kappa** should have the value of 1. However, for dimensional reasons, to agree with the $\mathbf{1/\Phi}$ it replaces, **Kappa** must have the units of m/kg. So Kappa is

Kappa = 1 m/kg.

Since with the gravitation of negative mass we are essentially in virgin territory we need to be guided by experiment. Since we have no flying saucers in captivity, we have to rely on the testimony of eye witnesses.

This testimony tells us that flying saucers have no inertia and that they produce gravitational fields strong enough to lifts objects from the ground. It turns out, that if you do the appropriate calculations, using **Kappa =1 m/kg**, the spinning negative mass ring of a flying saucer produces a field just enough to overcome inertia. And that value of the field is exactly the same value to create a gravitational force by even a small flying saucer, making a close pass, to lift a car off the ground. Therefore our value for **Kappa** is firmly based an "experiment", that is, what has to pass for experiment with flying saucers, eye witness reports.

1) Bondi, H. “Negative Mass in General Relativity”, Reviews of
Modern Physics, Vol. 29, No.3, July 1957, pp. 423-428.

Hoffmann, Banesh. Negative Mass, Science Journal, April 1965, pp. 74-78.
2) Martin, J., Ranada, A., Tiemblo, A., "On Mach's principle: Inertia as Gravitation", HYPERLINK < qc/0703141v1arXiv:grqc/0703141v1,2007>.
3) A. Einstein, The Meaning of Relativity, Princeton Univ Press, Princeton,1953.
4) D. Sciama, "The Origins of Inertia", Monthly Notices of Royal Astronomical Society, Vol. 113, page 34.
5. O. Jefimenko, Electricity and Magnetism , Electret Scientific, Star City W.Va., 1991, page 358.
6. O. Jefimenko, Causality Electromagnetic Induction and Gravitation, Electret Scientific, Star City W.Va., 1992, page 108.
7. SOBEPS, Vague d'OVNI sur la Belgique, I II, SOBEPS, Bruxelles, 1994. I, chapter 6.
8. R. L. Forward, "Negative Matter Propulsion", J. Propulsion, VOL 6, No. 81, p. 38 (1988)
9. CUFOS, ***UFOCAT*** 2007, Center For UFO Studies , Chicago, 2009 (UFO searchable Access database with 209,000 entries.)
10. J. A. Hynek, The UFO Experience, Marlowe Co., New York, 1998 , Chapter 9.
11. C. Kittel, Introduction to Solid State Physics, Wiley, New York, 2005, page 152.
12. Hynek, J. Allen; Imbrogno, Philip; Pratt, Bob; Night Siege, Llewellyn, St. Paul, MN, 1998, page 117.

20 Why This Book May Not Be Numerological Hallucination

I began reading about UFOs just out of curiosity with absolutely no preconceptions. The ideas and conclusions in the book were in a sense forced upon me by eyewitness reports.

The central conclusion, that UFOs have negative mass, is a consequence of the fact that a UFO can lift a car off the road momentarily. For that to happen the UFO has to produce a powerful negative vector potential **A(-).** And a negative vector potential can only be produced by negative mass.

The second conclusion, that negative gravity acts more like electromagnetism than positive gravity, is again forced upon us by eye witness observations. The negative gravitational effects mentioned above: the EM effects, the levitation, the paralysis, can only be real if the negative gravitational fields are strong enough, that is palpable or measurable. If we apply normal positive gravitational field calculations, the integrals have to be multiplied by a factor of 10^{-29}. This makes them immeasurably small. Therefore the observed effects dictate that the integrals are multiplied by 1, that is, negative mass behaves like electromagnetism.

And there is a simple explanation for this. Negative gravity is like electromagnetism because there is no **Mach's Principle** for negative mass. There is no accumulation of a universal NEGATIVE scalar field just as there is no universal accumulation of an electromagnetic field.

I have no illusions that everything in this book is correct. I think of it as "zeroth order theory", or what engineers call "proof of concept". To test whether the ideas worked, I set an artificially high standard. I wanted to see whether the smallest flying saucer observed, approximately 30 feet in diameter, could produce all the behaviors and effects seen in ALL UFOs. Because of physical constraints it is difficult for a small saucer to produce that large negative vector potential A(-) required to defeat inertia and provide propulsion.

By and large the test worked. Calculations even for such a small saucer showed just about all the behaviors and effects could be reproduced, with the possible exception of being inertia free at speeds of 2,000 miles per hour. So I believe that the basic ideas annunciated in the book are probably correct. I believe even more so because the picture of UFOs that emerges is so consistent.

From the physics of gravity we see that on object traveling in space at 1 km/sec, or 2,200 mph, generates a positive vector potential of **A(+) = $1.5(10)^{15}$ kg sec/m^2.** To counter that and eliminate inertia produced by this potential, a UFO would have to generate a negative vector potential, **A(-),** of basically the same magnitude.

This number served as reference point for the entire investigation. Could the UFO manufacture an **A(-) of $1.5(10)^{15}$ kg sec/m^2** ?

From the consideration of the physical constraints involved, it is possible, but only barely so, to construct a spinning negative mass ring in a UFO that will create a negative vector potential of

A(-) = - $1.5(10)^{15}$ kg sec/m^2.

It turned out that a vector potential **A(-)** of this amount was sufficient to

1. lift a car off the ground on a close pass
2. create a bubble thousands of feet wide where there is no inertia when the UFO is stationary and causes EM effects
3. propel the UFO with a reasonably slight displacement of the negative mass ring, and cause pendulum or "falling leaf" motion.

While the **A(+) = 1.5(10)^{13}** figure came from considerations of eliminating inertia up to speed of 2,200 mph, in doing the calculations, this is the one criteria that it appears might not quite be met. This value of the vector potential creates an guaranteed inertia free region only 2 meters in diameter, while the flying saucer's central mass may be 6 meters wide. To guarantee the larger diameter, the vector potential would have to be 10 times greater, that is, a ring mass of 400 tons or rotation speed five times the speed of light. These are not possible.

The matter is complicated by computational difficulties. Close to the UFO the normal formulas for the vector potential break down. We are not talking of wildly wrong figures, only inaccuracies of about on order of magnitude. We can just say with certitude what happens to the fields very close to the UFO. All we can say is that it appear that all UFOs will be free of inertia up to a speed of 500 mph, and a saucer with a tiny 10 foot core to 1,100 mph. It is possible, that the UFO will inertia free at higher speeds but our calculations are too inaccurate to tell. There are be possibilities for higher speeds but they appear to be only for larger Big Black Triangle UFOs.

In addition, from studying photographs of UFO "lights", lights that were always on and cast no beam, we estimated that a UFO had to disperse **10 kilowatts** of energy, probably used for the spin of the negative gravity rings. Therefore energy for propulsion, evidence of which was never seen, had to be less, say **50 kilojoules**.

What kind of mass could be accelerated to 35g with only 50 kilojuoules? The answer was about 1 kg. Therefore the effective inertial mass of UFO in the bubble would be **1/1,000,000th of its ordinary mass.**

This inertia would have to come from the miniscule negative scalar potential of the negative mass itself. Estimating this scalar potential , we found that inertia from this negative mass was $1/1,000,000^{th}$ of normal inertia from the universe. The remarkable coincidence of the two calculations of the effective inertial mass may be an accident. It nevertheless shows that two different approaches lead essentially to the same result.

I have no objection to anyone questioning any of the calculation in this book. I would point out, however, it is very unlikely that any error found will destroy the thesis of the book like a house of cards. Remember that the basis of the book are the eyewitness reports. Even if all analysis and calculations were dismissed, the reports will still be there. They have to dealt with.

As a graduate student at MIT I got into the habit of doing little calculations on backs of envelopes as I was reading physics to check that the results discussed where actually numerically sensible. I was reading about UFOs in ***Night Siege*** recently and thought that some of mysteries might be explainable. I again made little calculations to test the ideas out. I fully expected the results to be off the wall. To my surprise they weren't, they were quite reasonable.. As I began to read the eyewitness reports, I again did little calculations. Each time I expected the calculations to blow the ideas out of the water – but they didn't.

This then went on for two years

In physics, when you write a paper, it is almost always some incremental addition to a line of inquiry. Then you have to explain, that a larger, more detailed calculation would yield more, or that some unknown quantities would have to be understood. In other words, you have to sell or justify the work (if only to yourself). It is not often, possibly with Einstein with both theories of Special and General Relativity, with Heisenberg and Schroedinger in the original Quantum theory, that the results are accurate and immediately applicable. It is a pretty rare event.

So I found the fact that I was getting non nonsensical results in the calculations very unusual. After about two years of this, I got the feeling that I may be onto something. I stopped worrying about whether the picture I was developing was nonsense. As a matter of fact, if some calculations did not come out exactly right, I thought that I was barking up the wrong tree, and a little re-thinking might cure it. And I was right. I'll give two examples.

I originally thought that the spun negative mass was in form of a disk. When I did the calculations, with limitations imposed by the laws of physics and the properties of matter, it turned out that the spun disk did not create a large enough vector potential **A(-),** only $\mathbf{(10)^{11}}$ **kg sec/m^2**.. I then realized that a disk was not the most efficient form to produce maximal vector potential because the center of the disk moved at a slower rate then the rim, and it was the velocity of the mass that generated the vector potential. It was obvious that most of the mass had to be out near the rim. So the mass had to be something like a ring (like a Frisbee) where almost all of it moved at the rim velocity, which even then had to be as high as one half the speed of light. With this adjustment the calculation worked. (And incidentally, the usual picture of a flying saucer shows a space for such a ring.)

Then I was trying to adapt Forward's ideas of anti-gravity drive to the UFO. Forward's approach used the "perverse" inertia of negative mass to provide the propulsion for a positive-negative mass system. By applying a force to the negative mass by a "spring", the pull of the spring would provide a force, which by Newton's Second law and "perverseness" of the negative mass, would propel the mass in the opposite direction, AWAY from the spring.

This system could be stuffed into a UFO, with spring replaced by magnetic forces, and great accelerations could be achieved with only centimeter shifts of the negative ring. But the "spring" force one would have to apply to achieve accelerations seen in UFOs would be over a thousand tons.

And that would be little hard to do, since the shift mechanism could not handle such stress.

I then realized that Forward's theory depended on the "perverse" force being generated by being in the ordinary universe dominated by the universal scalar potential **Φ**. But the UFO was in a negative gravity bubble, where the effects of the ordinary positive gravitation fields are excluded. So Forward's theory could not be applied.

It turned out that it the large negative vector potential, **A(-)**, that the UFOs spinning negative mass generated that was the source of the force the propel the craft, the same force that levitated cars briefly. And the force applied to the negative ring to shift it would be very small. It would be achieved by shifting the ring 10 cm in one tenth of a second.

So every time I was momentarily stymied, it turned out that I was on the wrong track, and in a few days the correct approach became apparent. And the right approach was always better, and added greater clarity to an emerging picture. Adjusting the calculations were leading me to the correct picture. I felt like an archeologist working for years on a dig on an ancient temple, where bit by bit, the temple emerged, whole, and entirely intact.

I do not have any particular intellectual or emotional attachment to anything in this book. I thought that some of the puzzling things about UFOs could be explained. I took a serious look at the eye witness reports, now 209,550 in number, in the ***UFOCAT*** alone, and they revealed a lot. I think I have found some answers to a puzzle. I leave it to others to critique the solutions, or if it stands, to make it as basis for further study of the subject.

CODA

Anybody wishing to "debunk" the thesis of this book, that UFOs have negative mass, would have a formidable problem. The reason is that any kind of levitation implies negative mass. The proof has the simplicity of high school geometry resulting in QED.

There are about 300 reports of levitation in the ***UFOCAT***. The lifting of cars off the road üs only a spectacular example. But any levitation, no matter how slight, proves that the UFO is exerting a lifting force. This force can not be electromagnetic because that would effect tissue and metal differently. Strong magnets can lift metals, but only at a distance of inches. So the force has to be gravitational.

To deny the validity of these reports one would have to posit that witnesses all met in a café in Paris to plot a gigantic hoax, or that they were all delusional, reporting the identical effect over a period of 50 years from four continents. Short of that, any actual levitation, no matter how slight leads inexorably to the following conclusion:

Suppose the force of levitation is an acceleration of only a .1g, or 1 m/sec^2.

1. The levitation can not be Newtonian gravity because then the UFO would have to have a mass of $1.5(10)^{12}$ kg to exert such a force. That's a mass of a 2 km wide asteroid situated 10 meters from the witness.
2. The force would have to come from a gravitational vector potential A, that is, gravity from motion.
3. We do not know how, and it may be impossible to create a POSITIVE vector potential **A(+),** especially of a great magnitude.
4. The only vector potential we know is the positive vector potential **A(+)** of inertia, which is part of our universe.

5. The positive vector potential **A(+)** of inertia is REACTIVE. It is in the opposite direction of the acceleration that causes it. It is not a lifting force.
6. For the vector potential to be levitating it must be NEGATIVE.
   The vector potential **A** and the mass density are proportional.
7. A negative vector potential **A(-)** can only come from gravitationally negative mass.
8. Therefore if the UFO causes levitation it must have negative mass.
9. It is physically possible to generate a powerful negative vector potential **A(-)** because negative mass can be spun at the high speed necessary. Positive mass can not be spun because inertial effects will tear it apart.
10. If the negative vector potential **A(-)** is calculated using positive mass physics, then the factor G/c^2 is applicable. This makes the calculated value ravishingly small. A calculation implying that negative mass behaves like electromagnetism makes the calculation come out correct.
11. Therefore even one instance of levitation, no matter how small, immediately and inexorable implies that UFOs have negative mass. QED.

ORIGINS

21 Escaping to Another Dimension

During the Hudson Valley and Belgian waves, big black triangle UFOs would cruise at walking speed over the countryside in utter silence. They would then speed up and disappear. Where did they disappear to?

Indeed, the authors of Night Siege also wondered where UFOs went during daytime.

Because of the Belgian F-16 encounters we know that UFOs reflect radar since the fighters were able to lock on briefly. During the waves UFOs appeared and cruised around daily, why did they not show up on radars constantly ?

But the question is deeper than that. Near space is one of the most scrutinized areas of real estate in the world. Multiple radars constantly scan near space, keeping track of thousands of pieces of space junk so it does not collide with other spacecraft.

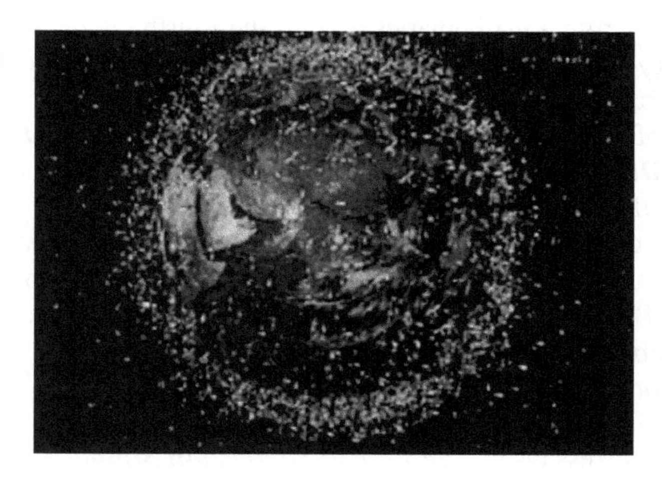

For decades the NORAD radars have been on hair-trigger alert for Russian missiles

Therefore it would seem that the million or so visits by UFOs coming over the horizon. during the last century did not come from outer space passing through that intensely scrutinized zone, for they then would constantly have triggered alerts.

It appears that UFOs show up on radar only at about the time they become visible, and only when they are already fairly low in the atmosphere. How come? The only answer is that they do not come from outer space. They come from another dimension where radar does not reach. Radar is, like light, electromagnetic radiation. If UFOs are not visible, they will also not appear on radar.

At about the turn of the Twentieth Century there was a great interest in four dimensional space, due mostly to the great influence of Einstein's Theory of Special Relativity. It was thought that a real fourth spatial dimension did not exist, because then all knots would unravel, and the inverse square law of gravity and electricity would not bind the moon in its orbit or the electrons around atoms.

A Professor of Mathematics at Brown University, H. P. Manning, wrote a mathematics text on the fourth dimension, Geometry of Four Dimensions (1). Scientific American held a contest for amateurs of essay about the fourth dimension. The volume of essays arising from the contest, published as ***The Fourth Dimension Simply Explained*** (2) in 1910, and it was edited and with an introduction by Prof. Manning. He established a tradition in the Brown Math department of work on the 4th dimension which exists to this day. Notably Prof. Thomas Banchoff has produces computer programs and films on the subject and the illustrated book Beyond the Third Dimension. (3)

In one of the essays in ***The Fourth Dimension Simply Explained*** the author describe what it would look like if a four dimensional object passed through our three dimensional space: (4)

> **...a four dimensional being moving steadily in the direction of the fourth-dimension might suddenly appear at our side within a room destitute of openings. Continuing this motion, the final limiting solid of this body would pass beyond our three-dimensional space into the fourth dimension, and he would disappear as suddenly and as inexplicably as he had appeared.**

In the Banchoff book there is an illustration of precisely this description. (5)

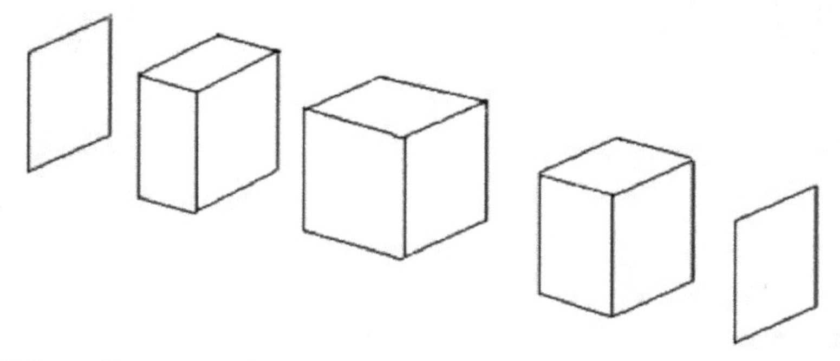

All four dimensional objects have parts which are three dimensional objects. A four dimensional square has 16 faces each of which is a solid three dimensional cube. A four dimensional hollow sphere appears as two solid 3-dimensinal tori (donuts) only one of which can be seen at one time.

As the 4-dimensional figure comes through our 3-dimensional space we can only see the 3-dimensional parts. Therefore the original statement about the 4-dimensional figure therefore also holds true for 3-dimensional ones. If a three dimensional figure came in from another dimension it would appear exactly like the description above, except when it was in our space it would be entirely in it, with no parts still invisible.

About half the authors in ***The Fourth Dimension Simply Explained*** try to analogize the relationship between three dimensions and a fourth by considering a fictitious two-dimensional world and how it would relate to three dimensions.

In that vein, suppose a two-dimensional world consisted of the surface of a lake as viewed from the bottom. One could then see the skis of a water skier moving on the surface of the lake.

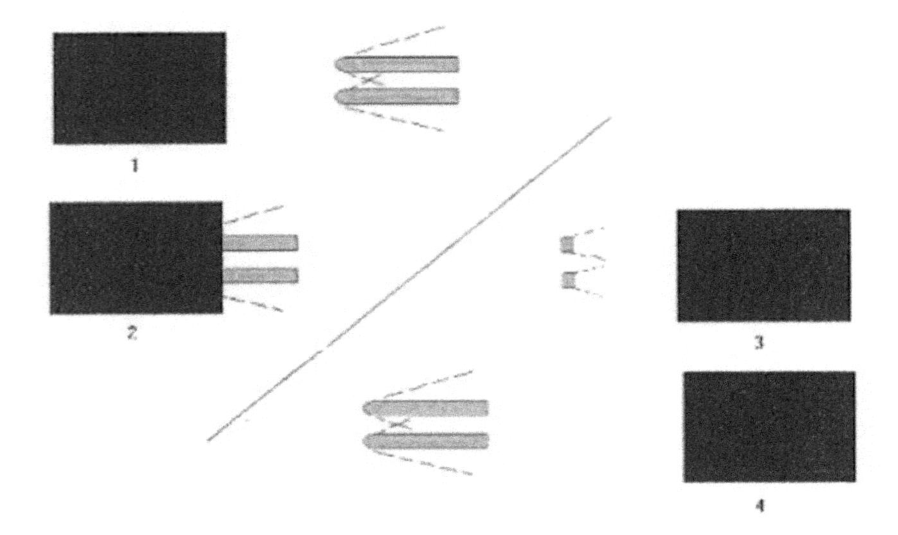

Suddenly the skier comes upon a large rectangle, and instead of crashing into it, the skier disappears, only to reappear on the other side. Of course we know that the water skier only came up on a ski ramp and used it to leap over to the other side. A viewer confined to two dimensions would think that the skier went through the block. As the skier flies through the air, the tips of his skis are elevated, and when he lands, the rear edge of the skis enter the surface first, and the rest of skis settle in the water.

What we have learned from this example is this: 1) entities able to enter another dimension appear to have the ability go through objects or walls, and 2) when an object enters from another dimension, it does not do so all at once, but gradually, like the skis settling in the water.

Let's know try to make a model of the 4-dimensional case. We will try to make this model as simple as possible. The real world model may be much more complicated.

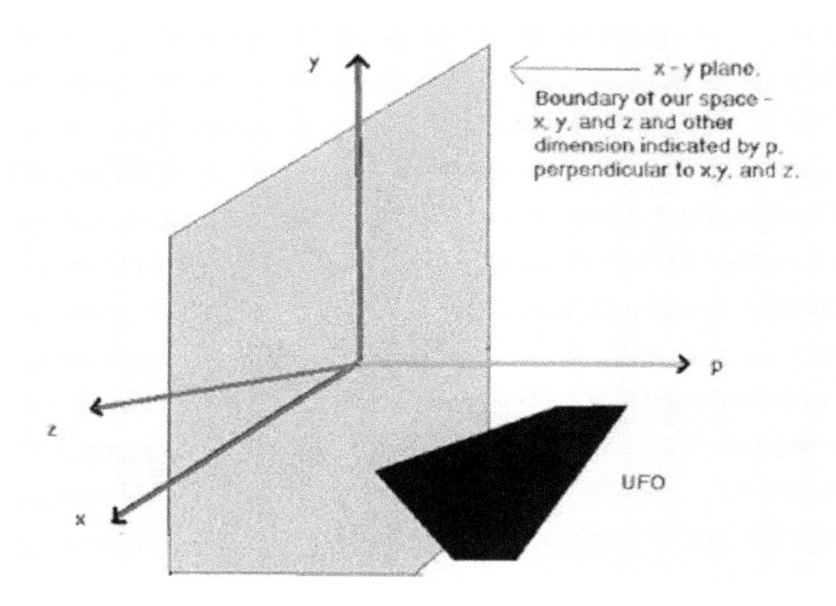

Suppose we have a 3-dimensional space with coordinates x, y, and z. Let us now add a fourth coordinate p, which is perpendicular to all the others. Let's say an object, say a UFO, is in the space x-y-p. Because the p direction is invisible to us, we can not see the UFO.

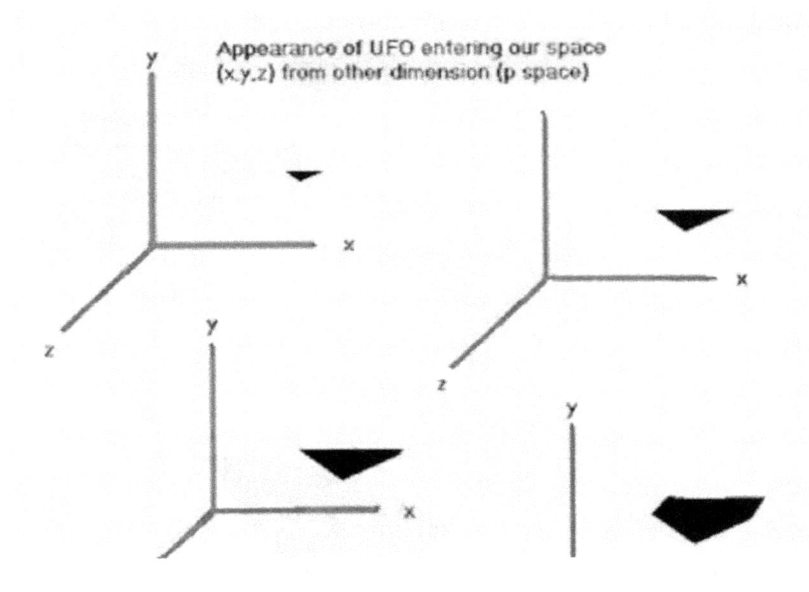

Lets now say the UFO approaches the origin of the p coordinate and then pokes its nose into our x-y-z space. Where the origin is and how the UFO does this, we do not know. A three dimensional object can only be in three dimensions at one time. Therefore the nose of the UFO is in x-y-z space, while the rest of it is in x-y-p space. You see that the entrance from another higher dimension into our x-y-z is piecemeal, and not all at once. As the UFO proceeds, it now is totally in x-y-z space and not in x-y-p space. The UFO now moves around in our world.

Why are we talking about UFOs? Because this is precisely what eye witnesses have reported UFOs doing. Over 10% of witnesses said that the UFO "shrunk" out of existence. (6)

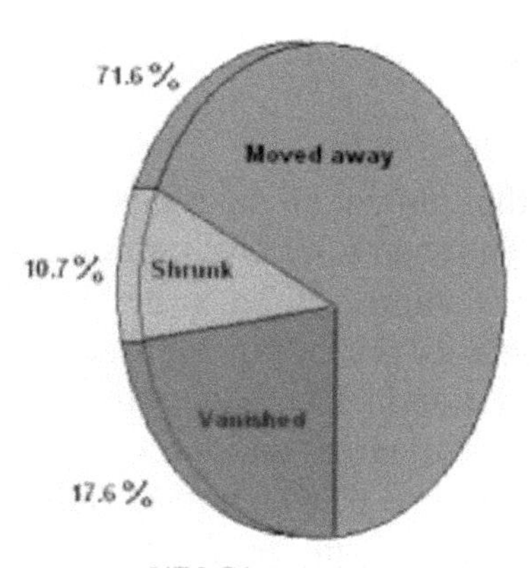

UFO Disappearance

We know of course that large material objects like a UFO don't "shrink". Their piecemeal disappearance at nighttime merely is interpreted by the eye as shrinkage. Witnesses said: " that the UFO, the size of a football field just collapsed on itself and vanished." They compared what they saw with "a huge telescoping antenna folding until it is very small." (7)

In another report: "The lights went out one by one, and, when the last light went out, the object had vanished like the Cheshire Cat's Smile." (8). The lights going out one by one is precisely what happens as the UFO withdraws into another dimension

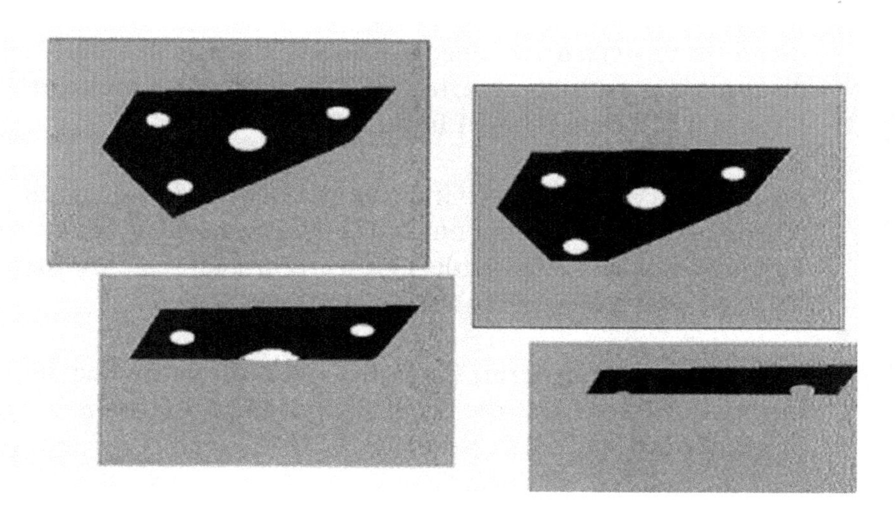

In another report said exactly that : **"The lights went out one by one."**

When you consider that almost all sightings are at night, it is very possible that even when a UFO is perceived to be moving away, that may be an illusion. When something gets smaller we unconsciously think it is receding because we are used to perspective. The reduction in size may actually be the apparent shrinking of the UFO as it enters another dimension, especially since the inevitable outcome of the motion is disappearance. The actually percentage of shrinking disappearance may be much higher than perceived.

During the night of April 29, 1990 when the Belgian F-16s made radar contact with the UFOs, the contacts usually lasted only a few seconds and the UFO would disappear from the screen only to reappear somewhere else a few moments later. (See Appendix A) Why would the contacts last only a few seconds? One could invent various scenarios where the UFO turned sidewise, or blocked the radar. Possibly.

But in the Hudson Valley wave a decade earlier the authors remarked:

"And there were many people who say the object simply disappear in an instant, often only to reappear a moment later in a blinding flash of light." (7)

An Officer Turnbull of the Greenburgh Police and his wife saw a UFO and the lights went out suddenly. It was **"as if the object had become invisible".** Officer Turnbull searched for the shape behind the lights but could not see it.

"About 40 seconds later, the lights suddenly came bac on, but in a different section of the sky not far from their original position."(8)

Since Officer could not see any shape were the lights had been perhaps we could conclude that the UFO DID disappear, and then suddenly reappear.

And that when the Belgian F-16s lost radar contact, it was not because of some radar ECM countermeasures, but that the UFO actually did disappear as in the Hudson Valley cases both from sight and radar.

UFOs, when they disappear do not go on long trajectories into outer space. Because they then could not be distinguished from an ascending ICBM missile to radars watching from half a world away. There is abundant information that UFOs are keenly aware of the situation on earth, and they would be loath to trigger a nuclear war. All of this confirms that UFO visit from another dimension, and return thereto.

UFOs seem to have the uncanny ability to pop into our universe, and then pop out again at will. It will be some time before we figure out how they do that.

According to recent work by the Harvard physicist Lisa Randall additional dimensions can exist which are infinite and not curled up, and they or may not have gravity. The problem with law of physics being violated as mentioned above then does not occur with additional dimensions.

The problem of whether a fourth or extra dimensions are physically possible, that we alluded to above, is dealt with in a subsequent chapter: "Extra Dimensions".

1) Manning, H. P., Geometry of Four Dimension, Dover, New York, 1956
2) 2) Manning, H. P., Fourth Dimension Explained, Munn, New York, 1910. (The above two tiles are available for download for free from Google Books.)
3) Banchoff, T. F., Beyond the Third Dimension, Scientific American Library, New York, 1990
4) Hynek, J. Allen; Imbrogno, Philip; Pratt, Bob; Night Siege, Llewellyn, St. Paul, MN, 1998, page 239
5) Ibid, page 239

6) Ibid, page 81.
7) Ibid, page 21
8) Ibid, page 28

22 Wormholes Not Required

Wormholes, or “space warp” as it is called in science fiction, where first mentioned by Herman Weyl in 1921 (1). Einstein in 1935 (2) discovered that there were mathematical solutions to the equations of general relativity that permitted two parts of the universe to be connected by shortcuts.

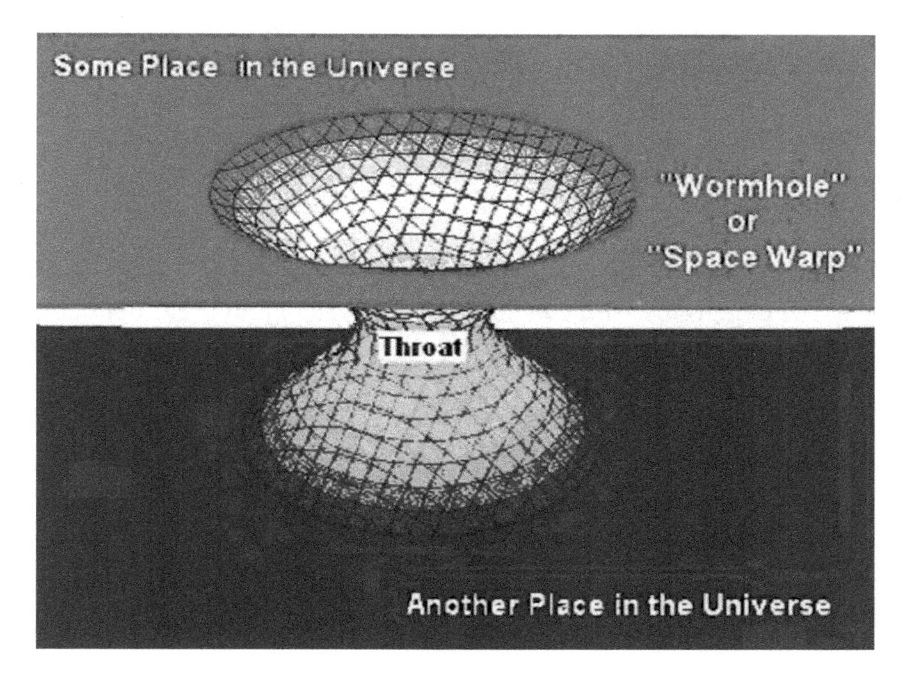

The “throats” of the wormhole were impossible to traverse in the original theories, as one encountered singularities and would be torn apart by quasi-infinite tidal forces. Later, Wheeler, coining the name “wormhole”, (3) discussed microscopic wormholes.

Finally in 1988 Kip S. Thorne (4) introduced traversable wormholes. The only catch was that to keep the throat open, one need “exotic matter”. “Exotic matter” is described as having negative energy density. But negative gravitational mass also provides negative space curvature, and could hold the throat open, but most likely would be needed in astronomical quantities, say a Black Hole. For some thirty years people have looked for evidence of such quantities of negative mass and not found it.

In the chapter called "Wormholes and Time Machines" in Black Holes and Time Warps" (5) Prof. Thorne describes the heroic efforts to get wormholes to become time machines, ultimately yielding to some objections by Steven Hawkins, and that the final answer will have to be eventually based on quantum gravity.

Lately there has been some discussion that this negative mass may not be needed, in Gauss-Bonnet type spaces where "mass without mass" solutions can be found with negative curvature. However, so far these ideas have not been developed further with respect to time travel. (6) In 1988 Matt Visser analyzed in a 412 page book, ***Lorentzian Wormholes*** (7), in great detail all of the wormhole theories to that date.

Although so far no one has seen any evidence that wormholes exist, nevertheless, over the years, a whole cottage industry has grow up around the subject in fiction. When Captain Kirk of the Enterprise says: "Scotty, warp speed!" he means going through a wormhole. If you would like to take a picture tour of what it would it would look like to traverse a theoretical wormhole, you can do so at

<http://www.spacetimetravel.org/wurmlochflug /wurmlochflug.html.>

As you see in the fanciful NASA illustration below, light (and gravity) extend through the wormhole and to the remote galaxy on the other side.

So there are several problems with "wormholes". So far they are wholly speculative. The argument is whether THEORETICALLY they exist. There is no definite answer to that question. Secondly no mechanism of how they are formed or kept open is even discussed. Wormholes means connecting one part of space-time to another with short cuts. It means that gravity exists and is obeyed at all times when one is dealing with wormholes. There is no clear answer as to what happens to objects as they traverse wormholes. There is no discussion that I know of the forces and energies involved. So at the moment wormholes are extremely theoretical speculations.

But what we know from eyewitness reports is that when a UFO enters another dimension, it is as if it where going through a mirror. (See "Escaping to Another Dimension") It disappears gradually. Parts that are already in the other dimension are invisible. Light does not follow the craft. Therefore UFOs when they disappear are not going into a wormhole, as in the NASA illustration. They are entering some space which is orthogonal to our space. In other words, while wormholes are attempted short cuts in space-time, UFOs seem to entering and traveling in regions which are not our space-time.

Since we do know that it appears that any theory that hopes to explain gravity and particles simultaneously requires that there be extra-dimensions, there are powerful physical reasons for such dimensions to exist. And as we know from the work of Dr. Randall, those dimensions could be infinite. Therefore time travel may be possible through those extra dimensions in ways we do not now understand.

1) Coleman, Korte, Hermann Weyl's Raum - Zeit - Materie and a
General Introduction to His Scientific Work, p. 199
2) Einstein, Albert and Rosen, Nathan. The Particle Problem in the General Theory of Relativity. Physical Review 48, 73 (1935).
3) Fuller, Robert W. and Wheeler, John A.. Causality and Multiply-Connected Space-Time. Physical Review 128, 919 (1962).
4) Morris, Michael S., Thorne, Kip S., and Yurtsever, Ulvi. Wormholes, Time Machines, and the Weak Energy Condition. Physical Review Letters 61, 1446–1449 (1988).
5) Thorne, Kip S., Wormholes and Time Warps, Norton, New York, 1994.
6) Gravanis, Elias and Wilson Steven, "'Mass without mass' from thin shells in Gauss-Bonnet gravity". gr-qc/0701152 (January 2007)
7) Visser, Matt, Lorentzian Wormholes, Springer-Verlag, New York, 1995.

23 Extra Dimensions

Physics today has two great very successful theories.

One it the theory of General Relativity. It was introduced by Einstein almost a century ago. It has withstood the test of time. It essentially explains the universe in the large, galaxies, planets, black holes and the origin and expansion of the universe. There have been many attempts at modification, none of which have taken hold.

The other theory is the Standard Model of elementary particles. It organizes the zoo hundreds of known particles and gives rational explanations for the relationship between them. It is not a complete theory since it cannot predict the absolute masses of particles, but it can show relationship between the masses. The Standard Model is actually the culmination of many other theories upon which it is based: quantum theory, quantum electrodynamics, gauge theory. The original quantum theory correctly described atoms. Quantum electrodynamics, which is known to be an approximation, can calculate some physical quantities accurately to 11 decimal places. All in all a very satisfying picture.

There is only one fly in the ointment. The two great theories are incompatible.

For fifty years people tried to "quantize gravity" and failed. They realized about twenty years ago that it couldn't be done. The reasons are deep in the mathematical structure of the theories.

As people realized that the two great theories could not be made to harmonize, they looked to additional dimensions to solve the problem. In 1921 Kaluza and in 1926 Klein (1) introduced the notion of extra dimensions. They were trying to combine electrodynamics with gravity and introduced a fifth dimension.

Today just about every serious theory that tries to unite particles and gravity, Supersymmetry, and Super String Theory etc., relies on extra dimensions. String theory started out with 35 dimensions, which by now has been whittled down to 11.

Quantization of gravity is needed to explain the gravitation force in detail. In quantum theories forces are mediated by spin one particles. We know that the spin of the gravity quantum, the graviton, is 2, but that is all we know. The trouble is that the world of quantum gravity is thought to exist at a dimension at what is called the Planck length, and that length is exceedingly small. It is 10^{-35} meters, 20 orders of magnitude smaller than the smallest particles we know of, the quarks.

There is no hope of doing experiments at that length in the foreseeable future. The only hope is some theory, and progress is slow. String theory has been in vogue for some time, but no none knows how to calculate with it, and many physicists, some of Nobel Prize caliber, are skeptical (for whatever that's worth.)

So at this point the only thing that seems evident is that any successful theory to combine gravity and quantum mechanics will have to incorporate additional dimensions.

But extra dimensions create problems for the law of physics. It was known already a hundred years ago that if you had a fourth spatial dimension, the lines of force of gravity and electromagnetism would have to spread themselves over this additional dimension, and the usual inverse square law of their fields would no longer hold.

Therefore it always had to be assumed that these extra dimensions curled themselves up on microscopically so that the field lines of gravity and electromagnetism could not spread out in them. This is called compactification.

Prof. Lisa Randall of Harvard was considering some of the

important unsolved problems in particle physics, like the hierarchy problem, and thought that extra dimensions might shed some light on the subject. The hierarchy problem has many aspects, but a main one is the question why gravity is weak, some 41 orders of magnitude weaker than electromagnetism.

In terminology adopted from string theory, our four dimensional world is considered to lie on a "brane", a two dimensional "membrane", which represents the four dimensional world.

In a paper, now dubbed RS1, with her collaborator Raman Sundrum, she was investigating the notion of two barnes, one the Gravity Brane, home to gravity, and the other the Particle Brane, our brane (1). The two branes would be separated in another or fifth dimension. But because they were only microscopically separated in the fifth dimension, we could not see the separation. The branes would appear to be superimposed. Here gravity could be strong on the Gravity Brane but weak when it got to our brane.

It was her assumption that this second brane was needed to terminate the effects of the first, and that the two branes were separated only microscopically in the fifth dimension by 10 Planck lengths. Then it was realized that the space between the branes was highly warped, hence the title of the book, Warped Passages. Warped in this case is a highly technical concept. The warping occurred because the branes had energy, and in general relativity energy curves (warps) space.

In the multidimensional world it means that the force of gravity falls off precipitously between the branes. She believed at the time that second brane was needed to terminate the extent of gravity to keep it contained in the microscopic, hence unobservable, space between the branes.

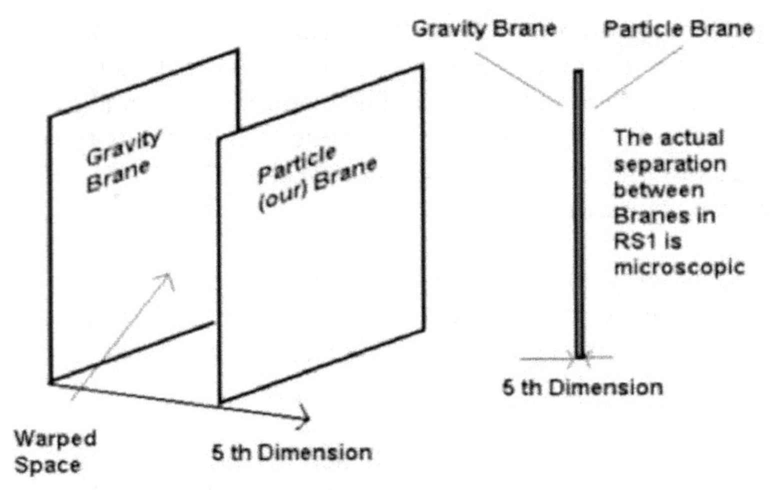

But on further calculation she discovered that the location of the second brane made no difference to the distribution of strength of gravity. Because of warping, the strength of gravity fell off precipitously by itself. The second brane could be anywhere, including infinity. In fact the second brane needed not to exist at all.

She and her co-author wrote this up in a second paper, now known as RS2, called "An alternative to compactification" (i.e. curling up) of the extra dimension (2). It turned out the gravity fell off so rapidly that little of leaked into the extra dimension. Therefore the laws of physics in our of dimensions still looked the same, even though an extra, a fifth, dimension existed.

Dr. Randall arrived at her conclusion from considerations of important unsolved problems in physics. But if we extract her conclusion form those considerations, what it says is that physics puts no barrier in the way extra dimensions existing, which can even have infinite extent.

In essence what this says is that the notion of an extra dimension we introduced earlier, the dimension p, is physically possible and does not create problems with the laws of physics. (See "Escaping to Anther Dimension".)

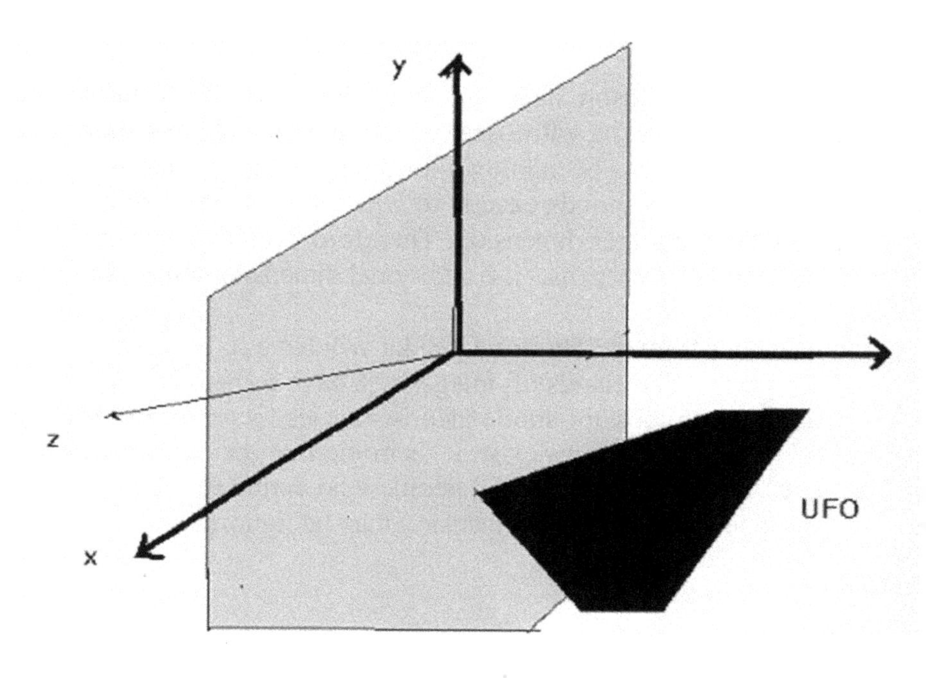

What is interesting is that in Dr. Randall's fifth dimension, gravity exists only near the brane (our world) but is absent in the rest of the (possibly infinite) dimension. Dr. Randall speculates that dimensions also exist where there is no gravity (4).

I have the sneaking suspicion that may be precisely why, since UFOs have been able to decouple themselves from gravity, to be able not care whether a part of the universe has gravity or not, that they are able to enter and travel into such dimensions, at will, as well as ours.

An Ironic Note:

In the introduction to Warped Passages, Dr. Randall says that she now believes that extra dimensions exist (3). She is awaiting for results from the new Hadron Collider particle accelerator which recently went on line that can reach higher energies to see if tell-tale footprints of additional dimensions begin to appear in the data.

During the Hudson Valley wave hundred saw UFOs shrink and disappear (5). The witnesses did not understand what they were seeing, but it has been known for a hundred years that what they saw corresponds exactly to what objects look like when they enter another dimension. Therefore from this eyewitness data one can infer that the additional dimension must exit.

The data from the Hadron Collider will have to be examined for years. Thousands of images will have to be minutely analyzed to look for subtle inconsistencies that indicate that additional dimensions exist. It is ironic that the evidence from eyewitness accounts of 700 people who actual saw UFOs disappear into another dimension may be ignored or dismissed.

1) Randall, L., Sundrum, R., "Large mass hierarchy from small extra Dimension", Phys. Rev. Letters, 83, 3370 (1999)

5) Randall, L., Sundrum, R.,"An alternative to compactification", Phys. Rev. Letters, 83, 4690 (1999)

2) Randall, Lisa, ***Warped Passages***, Harper, New York, 2006.

4) Randall, L. "Extra Dimensions and Warped Geometries", Science, Vol.296, Page 1422, (2002)

5) Hynek, J. Allen; Imbrogno, Philip; Pratt, Bob; ***Night Siege,*** Llewellyn, St. Paul, MN, 1998, page 239.

24 UFOs R Us

UFOs are not from a distant galaxy. They are terrestrial, most likely visitations by our descendants from the future.

You pick up a book on UFOs like ***Night Siege***, about the 1983-1985 UFO wave in the Hudson Valley, and settle back with visions of ET in the back of your mind. Then suddenly you are brought up short: some of the big black triangle UFOs that float at a walking place over the countryside have powerful steerable searchlights, somewhat like police helicopters, using white light. (1) Why WHITE light? Even animals here on Earth can see in ultraviolet and infrared. Why does ET prefer white light - like humans?

A little bit later on, an eyewitness trying to describe a particularly large UFO says it was the size of an aircraft carrier (2). UFOs are from 40' to 50' in size, going up to hundreds of feet, and then as big as a carrier. Why are UFOs exactly the same size as the (nautical) ships we build? Why are UFOs not the size of teacups or asteroids?

Finally, you read that UFOs are fascinated by lakes and reservoirs and spend a great deal of time hovering over them. Some even have arm like structures, probes, that descend from the hovering craft to examine the waters (3). Now think about how many planets such an instrument would be useful for. Mars has no lakes. Venus' surface is a rocky 900 degrees. Mercury is bare. Jupiter and Saturn are gaseous giants without a solid surface. So if you were to go exploring across the universe, would you build a special device into your craft - just to explore lakes - when the actual number of planets with lakes is one? Namely Earth. One begins to wonder.

So if we try to understand where UFOs are from, we can go two ways. First: We can look for a simple elegant answer based almost totally on what we know to be fact. Second: We can pile on hypothesis on hypothesis of things we have little knowledge of and also create a complicated picture to answer the question.

For those interested in understanding how statistically remote the possibility of intelligent life on planets is, we briefly review what we know about the evolution of life.

The Evolution of Life

We know the elementary particles contain within them the blueprint for inanimate matter. Elements are cooked in the 10 million degree cores of Type II stars. Protons and neutrons are assembled into the nuclei of all the elements, and as the nuclei migrate toward the edge of the star, they pick up electrons to form the elements. We know this because we can look all the way across the universe and way back in time see the in the spectra of stars that they are made of the same elements everywhere.

But when it comes to life there is no blueprint. If in the laboratory we combine the various elements that existed on the primitive Earth, water, carbon, nitrogen, etc. and subject them to an electrical discharge to simulate lightning, then some simple organic molecules that are precursors of life will form, but that's all. The process will not continue further.

To form more complex organic molecules one needs enzymes, and these enzymes have to come from some genetic system of DNA or RNA. But we have not been able to duplicate this step in the laboratory. And there is a good reason why. Because the blueprint in the elementary particles stops at this point, the next step has to occur by chance. And to say that this step is exceedingly rare is to put it mildly.

The Earth is about 4.5 billion years old. The fossil record of life dates back only to 600 million years. This means that for the first four billion years the Earth was lifeless. There are about 8 million lighting strikes on Earth per day, 2 million hit the ground, the rest stay in the clouds. If we assume that lighting is the catalyst of molecule formation, then with the rest of the material readily available, the beginning of life sure took its time. Its as if one bought 8 million lottery tickets per day, every day, and then had to wait 4 billion years, a good part of the age of the universe, to get a hit. You must admit these are exceedingly long odds.

Once life did form, it branched out. We have catalogued 2 million species so far, and there are estimates that there are 5 to 20 million species in total. Each species is a successful distinct example of life, being able to propagate itself.

What we need to understand is that the determinants of life are accident and history. Each new species appears the result of many accidental attempts at variation, one that succeed and was able to survive. Arguments about intelligent design or creation are philosophical and religious. A physicist is merely trying to find mechanisms that cause a result. So far no particular mechanism for speciation has been found, and the results are consist that it happens by chance.

But the 5 million species are not independent. They all grew out of precursor species so that's why biology is also history. A correct metaphor is the "tree of life". Each leaf on the tree represents a successful species. But each of these species is connected to twig, which connects to branch, and then to a limb of the tree, which are the sources of which species emerged from.

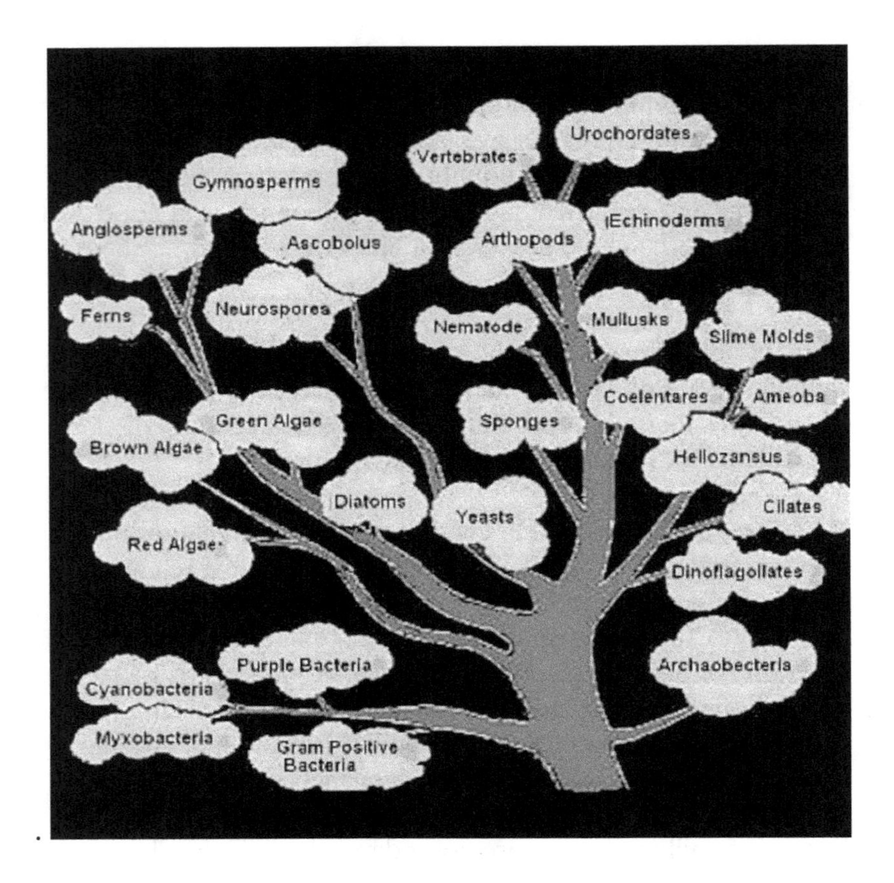

.

One leaf of the 5 million on the tree of life represents man. Its a very recent leaf, modern man having emerged only some 50,000 years ago, give or take. This means that a vertebrate bipedal toolmaker with intelligence is one chance result from 5 million successful experiments. Some guiding force for evolution can be seen. One is "convergence". To fill an ecological niche sometimes the same function arises from different sources, such as the wings of birds and bats, birds coming from dinosaurs and bats from rodents, each being different in structure, yet performing the same function.

Now suppose we say that UFOs are from a distant galaxy, or even somewhere from our own.

The first few hypotheses are based on some knowledge: there

are about 300 billion stars in our galaxy. Each one is thought to have a dozen planets. One of these planets may be "Earthlike". What that mostly means is the size, because the size determines the nature of an atmosphere.

If the planet is too large, its gravity will trap all the gases and be a mostly gaseous giant like Jupiter and Saturn in our galaxy. Of the three Earthlike planets, Mars, Venus, and Earth, both Mars and Venus are slightly smaller than Earth, but have very different atmospheres.

The atmosphere on Mars is about one hundredth of ours, while that Venus is a hundred times ours. The Venus atmosphere is mostly carbon dioxide and the green house effect traps heat so effectively that the surface temperature is a bout 700 degrees. The atmosphere is a soup and all the water is boiled off. Saturn's moon Titan has an atmosphere and photographs beamed of it surface look startlingly Earthlike, but its atmosphere and lakes are mostly methane. Life on other planets could be based on elements other than carbon, but we just don't have any knowledge of these.

So whether UFOs are from outer space, the ET hypothesis, would at first depend on the existence of a planet with a suitable atmosphere. We do not KNOW for a fact that such a planet exists, though they are plausible.

The next question is, does the planet have a chemistry from which some form of life can spring?

If so, will that giant statistical leap from molecules to life, that took 4 billion years on Earth, take place? We have no idea, since it is difficult to draw statistical conclusions from one example. If life on this hypothetical planet does form, will that life blossom into a "tree of life", or will it remain a "shrub" of life, with only algae and mud slime as its branches? Since the only life we know is accident and history, we simply don’t know.

We know that on Earth our species, Man, is only one of about 5 million successful attempts, that is, extant species. Something like that would have to happen on the planet on which ET evolved. "Little green men" appear to have a head, four limbs and a spine. On Earth man belong to the subphylum vertebrae, under the phylum chordates, of which there are 38 alone in the Animal Kingdom so on ET's planet such biological complexity must also exist if evolution works the same way as on Earth. And at this point, we have no reason to think otherwise.

If intelligent life forms do arise, on ET's planet, will they be toolmakers that can build UFOs? We have plenty of examples of fairly intelligent life on Earth, like giant squid and dolphins, but they don't use hammers and screwdrivers.

And finally, is there some evolutionary force of convergence that would make this tool maker develop legs, a head, and two eyes so that it would look like ET, like "little green men" - or us? We do not know.

Therefore the theory that the occupants of the UFOs are ETs which have evolved somewhere in the universe rests on a long chain of unknowns.

But then comes the biggest question. If such a being evolved somewhere in the galaxy, and set out to explore, would it find us, and if it did, would not it just see our relative state of underdevelopment and say "been there, done that" and move on?

Even our galaxy is a very big place. If a light beam left Earth at the time man first arose, 50,000 years ago, and traveling at 186,000 miles per second, it would only now have reached half way across the galaxy.

The question is, if a visitor from that far away were to visit us, would it hang around, …and hang around …and hang around?

The trouble is that there are 209,550 entries of UFO sightings in the ***UFOCAT*** debase alone. Since not all sightings are reported, some visits occur at hours of the night and places on Earth that don't report, the actual number of UFO visits is probably a million or more during the last century alone. Would extraterrestrial ETs hang around for a million visits?

But suppose we say the UFOs are not ET, but terrestrial.

A swarm of UFOs slowly cruised eastern Belgium for two years, another the Hudson Valley for three years. Why? Maybe because the Earth is their home planet, and Belgium and the Hudson Valley are their ancestral stomping grounds.

Then we can understand why they use white light to see: because they are human or humanoid, because they share our genes. They are our descendants. They see by white light.

They build the UFOs the size of our ships because they are the same size we are.

And they have probes built into their craft because their craft are built to explore Earth, and lakes on Earth are abundant.

So it would appear logically that the extra ordinary large number of UFO appearance are essentially visits by our descendants from the future.

The question is, is this possible? From all that we know about physics today, would physics allow time travel from the future?

Let us do a detailed analysis of this.

With respect to "wormholes" or "space warp" the situation is this:

1. While the solutions to Einstein's equations allow for such Wormholes, no one has put forth an actual model of how they would occur.It is said huge quantities of negative matter or energy would be needed to hold the

"throat" of such a wormhole open. Maybe something on the order of a black hole. People have looked for large quantities of negative and not found it.

2. Even so, time travel by such a possible wormhole is seems very doubtful. Kip S. Thorne and Stephen Hawkins have considered the matter and think some fundamental gravitational equations have to found before there is a solution.
3. Matt Visser, the author of ***Lorentzian Wormholes***, the definitive study, does not think time travel through wormholes is possible.

Then we come to extra dimensions. While wormholes imply that we go from one place in the universe to another by a short cut, extra dimensions mean that we actually leave our known universe.

1. Essentially all modern theories of elementary particles that want to encompass gravity require additional dimensions.
2. The additional dimensions were thought to have to be "compact" or curled up to make them consistent with laws of physics. New work by Dr. Randall of Harvard indicates that this may not be so. There could be dimensions with infinite extent leaving physical laws untouched.
3. We know from Special Relativity that time is part of a four-dimensional space-time continuum. Time is not fixed and absolute as Newton thought. Under certain transformations it is flexible. The "twin paradox", the premise of the ***Planet of the Apes*** movies is verified in the laboratory every day. If one goes fast, time slows down.
4. If additional dimensions exist, that means our space-time continuum is not four-dimensional, but many dimensional. But time is still one of the dimensions.
5. Since we know that even in our four-dimensional space-time, time is transformable, there is no reason to believe that it is not also so in a multidimensional

continuum.

6. Therefore we can not find anything in the physics that we know today that would definitely PROHIBIT time travel.

So it boils down to what is possible in a multidimensional continuum?

Unfortunately about that we know nothing at this point.

Dr. Randall herself has know come around to believe that extra dimensions exist. And then we have the evidence of hundreds of UFO eyewitnesses who saw UFOs “shrink” out of existence, the only rational explanation of which is that they entered another dimension. The technology of UFOs is more advanced than ours. Therefore, the conclusion that they are visits by our descendents form the future seems highly logical.

Dr. Jacques Vallee, the greatest living expert on UFOs came to this same conclusion years ago, for essentially the same reasons.

The occupants of UFOs seem to go to great lengths to avoid contact with us. But not quite. If you query ***UFOCAT*** database with the word "response", it returns a dozen instances where UFOs reacted to lights flashed at them by eyewitnesses, usually by flashing their lights back. So the people of the UFO seem to behave like humans. And in addition, handling their advanced technology, you would suspect they had intelligence and wit.

Their cruises of the Hudson Valley and Belgium did not seem to have a pattern, but appeared random and haphazard. One has the feeling the UFOs where tourists in very expensive space yachts curiously checking out how their ancestors lived centuries before.

There appeared to be only one personal encounter in the UFO wave described in ***Night Siege***. It is from a single witness. For whatever it's worth:

" then saw this figure wearing some type of suit approaching the object. ....As I watched, this guy raised his hand as if to say goodbye and vanished in a flash of red light. The object then vanished like someone had just turned it off. " (4)

The individual seemed to be our size, seemed to scamper around quite happily in Earth's gravity, and seemed to breathe our air. What really gets me is the raised hand. Does ET have human hand gestures?

Sure sounds like family.

1) Hynek, J. Allen; Imbrogno, Philip; Pratt, Bob; Night Siege,
   Llewellyn, St. Paul, MN, 1998, page 40.

2) Ibid, page 34.

3) Ibid, page 3.

4) Ibid, page 203.

5) Randall, L. Warped Passages, Harper, 2006

Politics

25 THE OBVIOUS

In retrospect it is obvious that people in the know, really smart people, say like those who built the atom bomb, for example, have from the very beginning understood the nature of UFOs.

It was clear that no aircraft could make the high acceleration moves of a UFO. This would be clear to anyone involved with aviation design. As an example, the SR-71 spy plane when it was still flying, needed the territory of three states to make a U-turn at speed. UFOs do it instantaneously. People were aware of UFO zig-zag motion. Zig-zag motion on its face violates the principle of **Conservation of Momentum**. The only way that is avoided is if the UFOs has NO momentum, that is, it is inertia free. Not only that, but they make their maneuvers in total silence, using effectively no energy, since energy use would have to show up somehow by the **Second Law of Thermodynamics**.

Sophisticated people have understood that inertia was mediated by gravity. In his 1952 book Einstein said so, and Sciama showed in 1957 that **Mach's Principle** was right, that gravity from the "distant stars" was the cause of inertia. So it was clear The UFOs were somehow manipulating gravity.

Various parts of the government, NASA, the Air Force, had for some 20 years secret research projects, upon which they spent millions, to develop an **anti-gravity drive**. (1) Since we did not have a UFO in captivity to examine in the lab, the Air Force also had various programs, finally called project ***Blue Book***, to gather information from the public on UFOs. UFO information from any source was scarce.

If on the one hand you are gathering information on UFOs and on the other hand spending considerable sums on **anti-gravity research**, it does not take much to conclude that you think

UFOs are using **anti-gravity**, as the maneuvers they perform can not be done by regular air craft.

For many years Dr. J. Allen Hynek, the Father of UFOlogy, was a consultant to it and front man for Project ***Blue Bo***ok. Dr. Hynek was a real scientist, at one time Chairman of the Astrophysics Department at Northwestern University. He eventually complained bitterly that the Air Force was not leveling with the public (2). He tried to convince the government to do a real investigation of the UFO phenomenon with no success.

It was always apparent, and especially through project ***Blue Book***, that UFO meant us no harm. They were entirely peaceful. The only remotely hostile activity was in Brazil, on the Island of Corales in 1977 when some juvenile delinquents in flying saucers were zapping the hapless local population with painful rays (3). Indeed, in 1975 the Air Force closed down project ***Blue Book***, concluding that UFOs where not a security threat. And by the way, this was exactly at the same time that it stopped doing **anti-gravity research.**

It was understood that there was nothing one could do about the UFOs. There have been attempts at shooting them down. There was a recent report about that in the British press (4). That activity was as likely to be as successful as primitive people shooting at jets overhead with bows and arrows. Since UFOs were an inevitable harmless presence, it was decided early on to simply ignore them and to deny the reality of their existence. It was reported recently that Winston Churchill and Dwight Eisenhower agreed to suppress a report of wartime UFOs so as not to alarm the public.(5)

What's that saying about letting sleeping dogs lie?

1) Cook, Nick, The Hunt for Zero Point Energy. Random House, NewYork, 2001
2) Hynek, J. Allen, The UFO Experience. Marlowe, New

York (1972), Chapter 12.
3) http://www.ufocasebook.com/colares1977.html
4) United Press International, "Brits Shoot at UFOs, Ex-official Says", January 27, 2009
5) The Independent, Aug. 5, 2010

26 The Separation of UFO and State

The Founding Fathers, when they wrote the Constitution, very wisely included provisions for the separation of Church and State. They had learned the lessons of History, the continuing European religious wars of previous centuries, especially the Thirty Years War. Granted, those wars where not directly fought over doctrine, whether there is One or Three Persons in the Godhead, or the apocryphal "how many angels can dance on the head of pin", but between peoples of different religions who were looking for political gain or dominance. But there were religious tests where people of a particular religion were favored or forbidden to hold office. The Fathers were guided by the thinking of John Locke whose essay: "A Letter Concerning Toleration" (1698) spelled out the problem and counseled tolerance. He argued that matters of religion were subjects for the individual's conscience not of the "magistrate", that is, the state.

There are things that governments are supposed to do: provide for the common defense, preserve order, and mundane things like build roads, sewers, and deliver the mail. And there are things that governments are not competent to do. One thing is to pick one religion as being better than another. Religion appears to be a universal concern of mankind, and societies deal with it differently. Attempts by governments to pick one religion as better than another have resulted in confusion and conflict.

There are certain things that are the patrimony of mankind. It is not for government to decide who really wrote the plays of Shakespeare. First of all it can't. Second of all it doesn't make any difference. The works of Shakespeare are the province of thoughtful mankind, not of government. Neither is cosmology. The Church for centuries had preached that the earth was the center of the universe. It did so on an overly literal interpretation of scripture. When Galileo asserted that the earth moved around the sun, he was condemned and subjected to house arrest. There were always plenty of bright people around the Curia. They knew that Galileo was right, but the Church had painted itself into a corner, and it was just too embarrassing to admit it, and to have to adjust its teaching. So through its authority it tried to maintain the charade for centuries, until that finally became an embarrassment.

UFOs are real physical objects. F-16s have radars whose sole purpose is to locate aircraft and possibly shoot them down. If F-16 radars lied, they would be useless as weapons. In 1990 Belgian F-16 radar made contact with UFOs. The radar was recorded on video. Presumably the tapes still exist. The something that reflected the radar was real (see Appendix A). Governments have access to the video tapes.

They have plenty of information on UFOs. For twenty years they collected information from the public in ***Project Blue Book***. They eventually came to the conclusion that UFOs were not a threat. Therefore the only legitimate reason for governments to be involved with the UFO subject at all vanished.

Yet governments maintain, and echoed by a subservient media, that UFOs don't exist, or that they can be explained away, or the that they are the result of the fervent imagination of kooks, and popular interest is safely vented in science fiction and by Hollywood.

When governments do not want to deal with a subject, they have the timed-tested technique of declaring the subject unknowable, and any one who attempts to say something intelligent is dismissed as "a conspiracy theorist".

The situation is so bad that the greatest living expert on UFOs, Dr. Jaques Vallee, the author of a dozen books, including the seminal ***Passport to Magonia,*** and on whom the protagonist of the major Hollywood the Spielberg movie ***Close Encounters of Third Kind*** was based, called his last book ***The Forbidden***

Science, changed careers and quit the field all together.

Public discourse on UFOs has been taken over entirely by various people with all kinds of agendas, some of them quasi-religious. Academics will not touch the subject with a ten-foot pole for fear of losing credibility and funding. This suits the government just fine. The reason is simple. First of all governments are convinced that UFOs pose no security threat, so there is no real worry. Secondly, governments do not have many answers, and it would be embarrassing to admit it. So they just use their propaganda machine to declare that UFOs don't exit and hope the problem, if it doesn't go away (UFOs always seem to came back) will be safely under the rug.

But UFOs are real. Reflection on them reveals aspects and dimensions of our physical universe we never dreamed of. UFOs make revelations that are truly part of the patrimony of mankind's view of the world. It is not the job of governments to tell mankind what it can think and know.

So I have a Modest Proposal. Let governments bite the bullet on embarrassment (since that is really the only thing at stake), declare that UFOs are real but governments really do not have all or even some of the answers. Then let mankind deal with UFOs. Governments do not even have to fund any research. Governments do not supply funds to religions. People do that themselves.

Simply acknowledge the Separation of UFO and State.

Miscellany

27 Jay XVI Goes for a Spin.

Jay entered the hanger where he kept his toys, the latest model spacecraft. He said he was in show business only so he could afford them. He called the hanger his "garage". He walked past the sleek Black Triangles and went to the corner where he kept his sentimental favorite, a small old style saucer, one of the smallest ever made. That's probably why it was in his collection. He decided to take it for a spin this morning. The saucer was 30 feet in diameter and was tied down like all the rest so they wouldn't float away. Having equal amounts of positive and negative mass, they would have drifted away like unmoored boats. The saucer was round because the negative mass was all in a giant ring at the base of the saucer. The ring had to be as wide as possible to generate enough negative field. That's why the old saucers all had wide flanges at the bottom to hold the ring. Above and below the ring where the superconducting magnets that acted as the bearings for the ring when spinning.

Jay opened the cupola and climbed into the saucer. The seats and the controls where well worn. He first flipped the switch to start the production of liquid nitrogen for the magnets. Just as Jay I had to light the pilot light and get up steam in his Stanley, these old saucers required a number of preparatory steps. With enough liquid nitrogen, Jay energized the magnets, and he heard that pop as the large ring centered itself in the huge shallow can. Then he started the vacuum pumps to get as much air out of the ring's container so there would be minimal friction for the tremendous spin that was to follow.

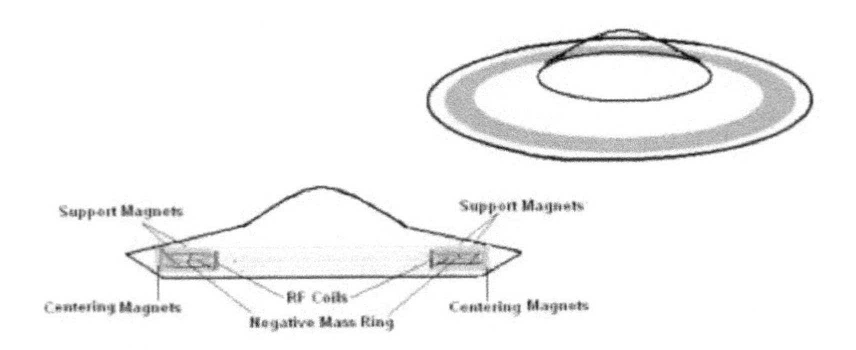

The auxiliary power supply was already plugged into the saucer, and Jay now turned it on to begin the slow process of spinning up the ring. The ring at this point weighed 40 tons and you did not have enough power in the saucer to begin the spin. The ring was spun by the magnetic coils around the edge of the can. In essence the coils and the ring formed an old fashioned AC induction motor. Later when the negative field killed the inertial mass of the ring, the coils could be powered by the saucer's own power supply. It took a little while to get the revolutions up. Slowly the negative field eliminated the ring's inertia and there was no longer the danger of centrifugal forces tearing the ring apart.

Finally the rotation speed reached nearly 5,000,000 revolutions per second and Jay switched to the saucer's own power. With ring's can evacuated and inertial mass now near zero you could hardly detect the motion. The driving coils around the edge of the can where now being fed RF at about 5 megacycles. The RF also excited the nitrogen gas creating a plasma around the saucer and made it glow in an intense pale blue-green. On take off the color would change to orange. This was the notorious bright glow seen by peoples all over the world for past centuries when flying saucers came to call.

One could not feel it, but Jay could tell from the instruments that thanks to the spinning ring which created a negative gravitational field which repelled the positive field of the universe, the saucer was now inside a bubble where there was no inertia. It is this which allowed saucers to make those incredible dashes and zig-zag motions that had always been the wonder of onlookers. It was also this lackof inertia which made the anti-gravitational propulsion work so well. With the saucer entirely free of gravitational forces, as you shifted the negative ring ever so slightly, the saucer would follow instantaneously. Ideas about this had been proposed by people like Herman Bondi and Robert Forward in the twentieth century, but nobody paid any attention.

Jay looked around the garage once more. Then he rolled back the roof of the hanger and tripped the tethers. He pulled on the stick. The servos lifted the ring just ever so slightly and the saucer popped up into the sky. Because the saucer now had no inertia, Jay did not feel the tremendous acceleration of the popup. The glow changed from green to orange as expected.

Jay sat for a moment looking at the Hollywood Hills below. It was a beautiful Sunday morning. He was not going to go far, like into another century. He pushed the stick forward. The servos moved the spinning ring ahead a bit, and the saucer shot off.

And there was a smile on Jay's face.

28 The Supplier's Problem

If you might want to build a flying saucer, I must just caution you, you there may be some problems. Suppose you went to your trusty supplier....

"Do you carry gravitationally negative mass"?

"Its on back order. We're out."

"I mean, did you have in the past?"

"I have eight other orders right now here in my computer."

"I need 40 tons of the stuff. Isn't that a lot?"

"No, actually some other orders are for much more."

"How much will it cost?"

"Can't say. All depends on circumstances when it comes in."

"When do you think that will be?"

"Don't know."

"Give me an idea. How long should I expect to wait?"

"As I said, I just can't say. Could be any day, or later."

"Can't you just take an educated guess? Make an estimate?"

"Ok, Ok! You are pushing me. Two hundred years."

"What ???"

"You wanted an estimate. So I gave you one."

“I don’t get it.”

“Look, we are in the supply business. It’s a business of supply and demand. You folks place orders, we supply. That’s how we make money. And we are not the bottle neck here.”

“Who is?”

“Its those guys at places like Cal Tech and MIT”

“How?”

“Look, we don’t know where to get the stuff. They won’t tell us.”

“Why them?”

“We don’t know if you have to mine it, or make it, or what.”

“What do you mean?”

“If you have to mine it, and it’s in a place like Tanzania, we can set it up. If you tell us Mongolia, we have connections there too.”

“Why mine?”

“If you tell us the stuff has to be made, we have Cracker Jack engineers in Taiwan. They will design a factory in no time, build it in China, and you will have the stuff in six months.”

“I still don’t get it.”

“Look, no one knows anything about the stuff. When it comes to the nitty-gritty of gravity everyone is clueless.”

“Why is that?”

“Look, those guys at Cal Tech and MIT, they write equations

for a living, right? Well they can't even write down a single equation about gravity, where it counts, what kind of charges there are and such. They keep chewing on their pencils, and we get nowhere."

"Why can't they write an equation?"

"Because everything is too damn small, that's why! They tell me smallest stuff they know is quarks, whatever they are. But the nitty-gritty of gravity, where the action is, is 20 orders of magnitude smaller. That's a decimal with a string of zeros 20 long before you get to a one. Pretty damn small, I tell you."

"Can't they do experiments?"

"Hah! That's why it will probably take 200 years before they figure out how to measure anything that small."

"So what happens?"

"Nothing. We sit and wait."

"God!

"Look, its pretty frustrating. We are in the supply and demand business. The only way we can make money is if we can come up with stuff that people want, and are willing to pay for. Like you."

"I am going to have to think if I need to change plans."

"Look, we are ready to go. My producers are ready, I am ready. We want to deal. We just have to wait for those others to get their ass in gear and tell us."

"Its kind of disappointing."

"You just gotta have patience. I got you in the computer. The minute I hear anything, I'll let you know."

29 The Great Physics Lab In The Sky

The UFOs are great teachers, especially of physics. By observing we can learn a great deal. Every night during a wave when the UFOs appear, it is another session of the Great Physics Lab in the Sky.

Now, did you read the chapter in the text about today's lesson, and the write up in the lab manual?

"Er…"

Watch as the UFO cruises at low altitude. It moves very slowly. It makes no sound."

"Wow! Look at all them lights. Green, red, purple. And they flash on and off!"

Have you ever seen airplanes fly this slow and so quietly?

"Look, Ma, I can run underneath and keep up!"

Why do you think a UFO can float like that?

"I dunno."

Why can a balloon?

"Because it's lighter than air!"

And a UFO?

"Because it's lighter than air?"

I don't think so. But can balloons do this, take off at 500 miles an hour and disappear?

"Where did it go? Where is it?"

UFOs float because they have equal quantities of positive and negative mass that cancel out their fields and the earth's gravity has nothing to hold on to.

"No kidding!"

The next night the UFOs are back. They are zig-zagging across the sky, making elbow turns on a dime.

Have you ever seen planes make such turns?

"No. But when the Bijou was having a gala, the searchlights did that."

Good observation. What you are saying is that UFO are behaving as if they were light beams, as if they had no weight or mass.

"I am?"

The hardest thing to get the mind around is that when an obviously massive UFO is traveling at a thousand miles an hour, it has practically zero kinetic energy and zero momentum.

"Gee! Zero, huh?"

Because kinetic energy and momentum both involve inertial mass, and the inertial mass of the UFO is near zero due the negation of inertia.

"They should have more galas at the Bijou with searchlights."

What that means is that all those eye witnesses who said UFOs looked solid, seemed made of metal, blocked out the stars and reflected radar where right. But those who said that UFO movements resembled light beams, holograms, were also right. UFOs are massive solid objects that flit around weightlessly because they have little inertial mass.

"You don't say."

So all the eye witnesses whose reports seemed so contradictory were both right. They both accurately reported what they saw. They were not lunatics or dupes.

"Yeah? Well, that's cool man . . .I wonder what's on ESPN?"

Capt Kirk, I don't think they are quite ready to learn, not just yet.

30 Epiphanies

The world of UFOs is that of the mind. It is not that UFOs are not real physical objects, it is that possibilities of interacting with them are so limited, that they take on the character of something that can only be perceived and not otherwise experienced.

The older models of UFOs would create all sorts of bizarre effects on the ground that bedeviled eyewitness. The more recent big black triangles no longer even do that. They floated above cities and highways at low altitudes and at walking speeds in Belgium, and the electrical systems of cars beneath them didn't die, and people didn't experience paralysis or levitation.

Religion deals with the spiritual, the wholly immaterial. UFOs on the other hand occupy a middle ground. They are physical objects but so remote that they essentially can only be seen. They do not choose to interact with us, and attempts to shoot them down have been deflected by them with childish ease. It looks like UFOs are likely to continue to visit, and this regime of look but don't touch will be with us for some time.

For those who bother to think, UFOs introduce a stunning reality of our physical world of which we did not dream. In this world gravity has been neutralized, what we thought were the inflexible barriers of time have been breached, and UFOs pass in and out of dimensions which we only now realize have to exist from the necessities of physics.

Then we are struck by sudden realizations, epiphanies, as it dawns on us that that UFOs very existence presents implicit messages.

Since the only rational explanation for them is that they are our descendants from the future, a message is that there IS a future. That we will not destroy ourselves in some nuclear holocaust.

That the ugly trait that is endemic to us, this irresistible compulsion to oppress and kill the weak, will in time be bred out. Since the time of Eannatum of Sumer in 2,500 BC, at the beginning of recorded history, the story inevitably shows, that at least for those in power, there is a primitive urge to trample on the weak, and if they resist, to kill them.

There is no doubt that with their technology UFOs could do us great harm. Yet they have not done so. They have been with us, possibly for centuries, and the only remotely aggressive acts were in the island of Corales in Brazil in 1977. Their otherwise total peacefulness points to an eventual similar future for us.

And finally we realize that we are going to BECOME them, the beings of the UFO. That we will have the power, the exhilaration, the absolute freedom to defy gravity, breach the boundaries of time, and pass through other yet unseen dimensions. In our world gravity is like flypaper. It traps us. It makes us crawl around our solar system with unwieldy rockets that take years to reach another planet. It is highly likely that the UFOs ability to enter other dimensions is related to their ability to decouple from gravity.

The poem "High Flight" has become something of an anthem for astronauts.

Oh! I have slipped the surly bonds of Earth
And danced the skies on laughter-silvered wings.....

John Gillespie Magee, Jr

When we become THEM, the people of the UFO, we will not only have slipped the surly bonds of Earth, but the surly bonds of gravity, and the bonds that bind us to our little corner of the universe.

Appendix A

The following is a report written by the secretarial staff of the Belgian Air Force.

REPORT ON THE OBSERVATION OF UFOs DURING THE NIGHT OF MARCH 30-31, 1990

1.Introduction

a. This report gives an overall view of the reports from the concerned Air Force units and of the reports from ocular witnesses of the gendarmerie patrols, about the unknown phenomena watched in the air space (hereafter called UFOs), south of the axis Brussels-Tirlemont, during the night of March 30-31, 1990.

b. The observations, visual and radar, were of such a nature that the take off of two F-16 of the 1 J Wing has been decided, in order to identify these UFOs.

c. This report has been established by Major Lambrechts, VS/3 Ctl-Met 1.

2. Context.

Since the beginning of December 1989, strange phenomena have been regularly noticed in the Belgian air space. The Air Force has at its disposal several ocular witnesses, most of them having been informed by the gendarmerie. The Air Force radar stations could not confirm, in any case, up to March 30-31, these sightings, and the presence of the UFOs could never be established by the fighters sent in that order. The Air Force staff has been able to produce several hypotheses about the origin of these UFOs. The presence or the testing of B-2 or F-117 A (stealth), RPV (Remotely Piloted Vehicles), ULM (Ultra Light Motorised) and AWACS in the Belgian air space during the facts can be excluded.

The cabinet of the MLV (Ministry of National Defense) has been informed about these discoveries. In the meantime, the SOBEPS (Societe Belge d'Etude des Phenomenes Spatiaux) got in touch with the MLV, in order that the MLV backed the SOBEPS in its inquiries about this phenomenon.

This request has been accepted, and after that the Air Force has regularly cooperated with this society.

3. Chronological summary of the events during the night of March 30-31, 1990. Note: local time.

March 30:

23 h 00: The supervisor responsible (MC) for the Glons CRC (Control Reporting Center) receives a phone call from Mr. A. Renkin, gendarmerie

MDL, who certifies to see, from his home at Ramillies, three unusual lights towards Thorembais-Gembloux. These lights are distinctly more intense than stars and planets, they don't move and are located at the apexes of an equilateral triangle. Their color is changing: red, green and yellow.

23 h 05: The Glons CRC asks the Wavre gendarmerie to send a patrol at this place in order to confirm this sighting.

23 h 10: A new call from Mr. Renkin points out a new phenomenon: three other lights move towards the first triangle. One of these lights is far brighter than the others. The Glons CRC observes in the meantime an unidentified radar contact, about 5 km north of the Beauvechain airport. The contacts moves at about 25 knots towards west.

23 h 28: A gendarmerie patrol including, among others, Captain Pinson, is on the premises and confirms Mr. Renkin's sightings. Captain Pinson describes the observed phenomenon as follows: the bright points have the dimension of a big

star(*); their color changes continually. The prevailing color is red; then it changes itself in blue, green, yellow and white, but not always in the same order. The lights are very clear, as if they were signals: this enables to distinguish them from stars.

23 h 30 - 23 h 45: The three new lights, in the meantime, have drawn closer to the first observed triangle. In their turn, after a series of erratic moves, they arrange themselves also in triangular formation.

In the mean time, the Glons CRC observes the phenomenon on radar.

23 h 49 - 23 h 59: The Semmerzake TCC/RP (Traffic Center Control/Reporting Post) confirms in its turn to have a clear radar contact at the same position pointed out by the Glons CRC.

23 h 56: After prerequisite coordination with the SOC II, and since all conditions are fulfilled to make the QRA take off, the Glons CRC gives the scramble order to the 1 J Wing.

23 h 45 - 00 h 15: The bright points are still clearly observed from ground. Their respective position does not change. The whole formation seems to move slowly in comparison with the stars. The ocular witnesses on ground notice that the UFOs send from time to time brief and more intense luminous signals. In the mean time, two weaker luminous points are observed towards Eghezee. Those, as the others, have also brief and erratic moves.

March 31:

00 h 05: Two F-16, QRA of J Wing, AL 17 and AL 23, take off.

Between 00 h 07 and 00 h 54, under control of the CRC, on the whole nine interception attempts have been undertaken by the fighters. The planes have had, several times, brief radar contacts on the targets designated by the CRC. In three cases, the pilots managed to lock on the target during a few seconds, which, each time, induced a drastic change in the comportment of the UFOs. In no case, the pilots have had a visual contact with the UFOs.

00 h 13: First lock on the target designated by the CRC. Position: "on the nose" 6 NM (Nautical Miles), 9000 feet, direction: 250. The target speed changes within minimum time from 150 to 970 knots, altitude coming down from 9000 to 5000 feet, then up to 11000 feet, and, shortly after, down to ground level. From this results a "break lock" after some seconds, the pilot losing the radar contact. The Glons radar informs, at the moment of the break lock, that the fighters are above the target position.

+/- 00 h 19 - 00 h 30: The Semmerzake TCC as well as the Glons CRC have lost contact with the target. From time to time a contact appears in the region, but they are too few to have a clear track. In the meantime, the pilots contact on VHF the radio of the civilian air traffic, in order to coordinate their moves with the Brussels TMA.

The radio contact on UHF is maintained with the Glons CRC.

00 h 30: AL 17 has a radar contact at 5000 feet, 20 NM away Beauvechain (Nivelles), position 255. The target moves at very high speed (740 knots). The lock on lasts during 6 seconds, and, at the break lock, the signal of a jamming appears on the scope.

+/- 00 h 30: The ground witnesses see three times the F-16 pass along. During the third pass, they see the planes turning in circles at the center of the great formation initially seen. At the same time, they notice the disappearance of the little triangle, while the brightest, western point of the big triangle moves very fast, probably up. This point emits intense red signals, in a repetitive way, during the maneuver. The two other points of the great triangle disappear shortly after. The clear points above Eghezee are no longer visible, and only the western brightest point of the triangle can be observed.

00 h 32: The Glons and Semmerzake radars have a contact at 110 / 6 NM away Beauvechain, which heads for Bierset at 7000 feet and high speed.

The registered speeds go from 478 to 690 knots. The contact is lost above Bierset. The Maastricht radar control center has had no contact with this UFO.

00 h 39 - 00 h 41: The Glons CRC mentions a possible contact at 10 NM from the planes, altitude 10000 feet. The pilots have a radar contact at 7 NM. Again is noticed an acceleration of the target from 100 to 600 knots. The lock on lasts only a few seconds, and the planes as well as the CRC lose the contact.

00 h 47: The Beauvechain RAPCON mentions a contact on its radar, at 6500 feet altitude, position away Beauvechain: 160 / 5 NM. The Glons CRC has also a contact on the same position. This one is observed up to 00 h 56.

00 h 45 - 01 h 00: Some attempts are undertaken in order to intercept the UFOs. The planes register only a few very short radar contacts.

The ground observers see the last UFO disappear towards Louvain-la-Neuve (NNW). Around 01 h 00, the UFO has completely disappeared.

01 h 02: AL 17 and AL 23 quit the frequency of the Glons CRC and go back to their base.

01 h 06: The Jodoigne gendarmerie mentions to the Glons CRC that has just been observed a phenomenon like the one observed by Mr. Renkin at 23 h 15.

01 h 10: Landing of AL 17.

01 h 16: Landing of AL 23.

01 h 18: Captain Pinson, who in the meantime has gone to the Jodoigne gendarmerie, describes his observation as follows: four luminous white points at the apexes of a square, the center of which is Jodoigne. The UFO seen towards Orp-Jauche (SW of Jodoigne) is the brightest and has a yellow-red color. The luminous points move with jerky and short moves.

+/- 01 h 30: The UFOs lose their luminosity and seem to disappear in four distinct directions.

4. General information.

a. Meteo. The data mentioned by the Air Force Wing Meteo regarding the concerned area and during the night of March 30-31, 1990, are the following:

Visibility: 8 to 15 km with clear sky. Wind at 10000 feet: 50/60 knots. A slight temperature inversion at ground, and another, as slight, at 3000 feet. These data are confirmed in Captain Pinson's report. He mentions also that the stars were clearly visible.

b. Because of lack of appropriate material, the ground observers could not make any photo or film of the phenomenon.

c. The UFO observed with a telescope is described as follows: a kind of sphere, a part of which is very luminous; a triangular shape could also be distinguished (For a more detailed observation, see Captain Pinson's report, in appendix H1).

5. Constatations.

a. In contradiction with other pointed out UFO sightings, for the first time a radar contact has been positively observed, in correlation with different sensors of the Air Force (CRC, TCC, RAPCON, EBBE and F-16 radar), and this in the same area as visual observations. This has to be explained by the fact that the March 30-31 UFOs have been noticed at +/- 10000 feet altitude, whereas in the former cases there was always talk of visual contacts at very low altitude.

b. The visual evidences, on which this report is partially based, come from gendarmes in duty, whose objectivity cannot be questioned.

c. The UFOs, as soon as seen by the F-16 radar in the "Target Track" mode (after interception), have drastically changed their parameters.

The speeds measured at that time and the altitude shifts exclude the hypothesis according to which planes could be mistaken for the observed UFOs. The slow moves during the other phases differ also from the moves of planes.

d. The fighter pilots never have had visual contact with the UFOs. This can be explained by the changes of luminous intensity, and even the disappearance of the UFOs, when the F-16 arrived in the neighborhood of the place where they were observed from the ground.

e. The hypothesis according to which it was an optical illusion, a mistake for planets, or any other meteorological phenomenon, is in contradiction with the radar observations, especially the 10000 feet altitude and the geometrical position of the UFOs between themselves. The geometrical formation tends to prove a program.

f. The first observation of the slow motion of the UFOs has been made roughly in the same direction and with the same speed as the wind.

The direction differs by 30 degrees from the direction of the wind (260 degrees instead of 230 degrees). The hypothesis of sounding balloons is very improbable. The UFOs altitude during all this phase remained 10000 feet, whereas the sounding balloons go on higher and higher, up to burst at around 100000 feet. It is difficult to explain the bright lights and changes of color with such balloons.

It is very improbable that balloons stay at the same altitude during more than one hour, while keeping the same position between themselves. In Belgium, during the radar observation, there was no meteorological inversion in progress. The hypothesis according to which it could be other balloons must be absolutely dismissed.

g. Though speeds greater than the sound barrier have been measured several times, not any bang has been noticed. Here also, no explanation can be given.

h. Though the different ground witnesses have effectively pointed out eight points in the sky, the radars have registered only one contact at the same time. The points have been seen at a distance one from another sufficient for them to be distinguished by the radars also.

No plausible explanation can be put forward.

i. The hypothesis of air phenomena resulting from projection of holograms(*) must be excluded too: the laser projectors should have been normally observed by the pilots on flight. Moreover, the hologram cannot be detected by radar, and a laser projection can be seen only if there is a screen, like clouds for example. Here, the sky was clear, and there was no significant temperature inversion.

(end of report)

(*)sic

This is the transcript of the exchanges between the pilots and the ground controller during the interception of "UFOs" over Belgium, night 30-31 March 1990 (TIME is GMT).

30/03/90 QRA(I) SCRAMBLE

TIME C P ITEMS TAD 465

2207
C Loud & clear how me
P Reading you 5, flight level 90
C Task VID check armament safe
P Safe
C For your info, contact at your bearing 310 range is 15
P 310, 15, and confirm it's still on FL 90
C checking

220730
P Bravo reading you 5
C Bravo 5 as well. No
C No height on the contact for the moment
P Both leveling off FL 90
C Roger and both starboard 310
P OK, SB 310
C Last altitude on the contact is FL 210
C Keep on turning, roll out 320
P 320

2208
C 320, 17 miles. And for the moment maximum level 10000ft
P Steady 320
C Roger 330, 5 to 10 right range is 15
  Possible altitude 10000ft
P Steady at 10000. No contact

2209
C Contact 330 range 10. 11000ft
  Starboard 330
P Steady 330
C 330. 5 right range is 9
P No contact keep on taking
C 345 range 7. Reduce speed. Slow moving
P Roger, slow moving
C Still at 10000ft. Bearing 345 range 5

P Confirm altitude

2210
C Last altitude 10000ft. Check 10 left range is 3. Left side 2 miles.
No altitude. Passing overhead
P No contact
C Just below you
P Say again
C Just below you now. Both vector 090.Contact is 090 range 2
When steady check 090 range 3. Slow moving.

Inside turn 4 Nm, 060, 3.
P One blinking light just in front of you, do you see it, just below
you. An orange one.
C Range is 3, 060, 3.
P ... heading 180. Roger reversing 180 You have contact on me MEEL. Roger contact on you.
C If you reverse 180, on your nose 1 mile. It should be 1 o'clock for you. Blinking orange light. It's on the ground P Efflux, you still have the contact.
C Contact for the moment 020, 15
P Confirm 020
C 020, 5 miles
P See the blinking light I mean unreadable....flash
C 030, 6 miles
P Contact on the ground seems to be 1 light
C Another contact now 360, 10 miles
P 360, 10

2213
C Altitude 11000ft, 350, 11 miles
P I have a contact 9000 heading 250 at 970 knots
C Possibly your target
P One contact on the nose 9000ft speed 310
C Range is 6?
P Eddy do you confirm contact. I have the same in
B 15 now unreadable

C Contact is at 3 miles now. On the nose 3
P Contact is coming in and out
C Roger and now... 2miles Right inside turn, level 1 mile

2214
Expedite right, roll out 130
P unreadable. 130
C 140 range 3
P Confirm heading efflux
C 130, 120 even. And continue roll out 180 He's now 170,4.
Check camera on. 160, 3.
P Camera on I've a possible contact now at 550 knots in C. 6
alt 10000
C Just overhead

2215 If possible take a maximum of pictures.
P May I suggest you keep the HUD, I keep...
C At your 6 o'clock 2
P unreadable
P Efflux, give a new heading
C Roll out 360, 360, 2. unreadable Continue SB 030
P 030
C He's now 050, 3. Altitude 105. Keep on turning towards 090
P Steady 090
C 090 on the nose 2
P One a/c passing below. Efflux is it possible ?
C At what altitude ?
P I see it efflux
C On the nose 2 miles
P MEEL, you see it, just below me now Efflux you have a new
heading
C South, 2
P Say altitude
C FL 105. Snap 130. 130, 3. Last alt. reported 10000ft
On the nose 2
P come in attack
C Past the contact now. Altitude is 10000ft
P I'm at 9000ft
C Still no contact?

P " " " ! Heading please
C 270, 2
P Confirm 260

2219
C 270

2220
P Roll out 270. Steady 270. 10000ft
C No more contact for the moment
P MEEL you switch 135 05 go
C Can you contact Brussels on 127.15
P 127.15, go Efflux, confirm new heading

2221
C Keep on turning right 090
P Turning left 090. Efflux steady east now
C Roger, maintain
P Positive contact as well
C For the moment no more contact on the scope

2222
P No contact on the scope as well Check fuel.
unreadable Possible contact at 19 miles 800 knots h 350

2223
3000ft // Efflux confirm one contact at 5 miles, left side, speed fast
C No contact for the moment
P 4 miles to the left
C Clear to investigate
P Investigating. Rolling out now 034

2224
Brussels is calling. No contact.

C Traffic approaching from 320 range 15, 9000ft.
Possible contact bearing 270 range 12.

Starboard turn

2225
P Turning right 270 This contact seems to be civilian traffic
P Say again Efflux
C Contact is civilian traffic

2226
P Rolling out 277
C Roger, maintain 17 from efflux
P Come in efflux
C Did you see in the previous investigation...
P I had a kind of flashing light on the nose 5 miles
C And this light was coming from the south?

2227
P This light is steady
C When did you pass over the light, give me a top
P Turning left to pass overhead at 10000ft and give you the coordinates Just passing overhead the light
C Roger
P Coordinates : 50.32.08 04.11.08 Reversing east, 10000ft
C Roger
C Possible contact bearing 020 12 miles
P 12 miles looking out
C High speed roll out 040
P 040
C Heading is 115 Starb 060
P One contact on nose 10 miles
C That's the target. No alt on him for the moment
P Contact in C 12 MEEL, at 5000ft. 740 knots Good contact again.
Investigating One contact on the nose 7 miles
C Clear to investigate, check armament safe

2231
P Sweet and safe
C Passing overhead BE for the moment
P Lost contact now, he's moving very fast

C That's affirm High speed for the moment
P One contact on the nose 6 miles, speed to 100 knots
C 080, 10 miles. Heading is 120
P 120 confirm
C Affirmative

2232
C Last alt. reported 10000ft 070, 10 miles
P 070, 10 confirm Rolling out 070. Altitude 7000ft Lost contact
more info efflux
C Lost contact as well. It should be 090, 10 Roll out 100

2233
P 100
C Normally on the nose range 15 You have contact

2234
P No contact
C 095 Range 18 17, both starboard 310
P SB 310 Fuel 044
C 17 check playtime left
P Playtime left 15 minutes
P 17 steady 310
C Roger 17. Maintain hdg for the moment One civilian traffic 315 range is 12
at 5000ft in the TMA 2238
P Looking out Contact at 6000ft slow moving at C
C It's civilian traffic. Passing 2 o'clock 5 miles 5000ft, check 310, 12 miles
possible contact

2239
P 10 miles on the nose 10000ft Contact
C On the nose range 7 2240
P Got the same
C Check camera on 2241
P Camera on
C If possible take max of pictures

P Very slow moving
C Check alt. of contact
P I still have the contact, 5 miles
C No height
P No height
C 3 o'clock 2 miles
P " " "
C Crossing left to right
P Say again
C Left side high
P Looking out. I see one beacon on the nose 2242
C One civilian traffic west, 10 miles
C Contact 100, SB 100
P Roger SB 100
C Civilian traffic 300, 5 miles
P " " " " "Steady 120
C Continue 100
P 100
C Even 060 now 060, 5
P Steady 060
C 060, 3. You have contact?
P One contact but speed is changing from 100 to 600
C I have the same contact
P Slightly to right 4 miles
C Affirmative. High moving
P Steady east now
C Roger
P Lost contact
C Both vector 180

2244
P Turning right south
C Contact south higher
P Looking out. Steady south.
C Nine o'clock 3. sorry 3 o'clock
P Steady south no contact 2245
C Disregard snap 360 now
P 360 to the left Check fuel
C Possible contact 350 range 10

P 350, 10
C 2 contacts due to civilian traffic same position 345,
9 left 330 17 left 330 left 330 2246
C Civilian traffic 340 range 7
P Contact on traffic
C At 5000ft, other contact at 325 range is 7, no height
P Contact on the radar now
C Check camera on
P Camera on loosing contact
C He's now 345 range is 5
P We have the same in B 8, 10000ft. MEEL
C 350, 3
P Radar contact slightly to the left, 8 miles, lost contact now

2247
C He's at your 360 now 360...
P Request to turn north
C Clear now
P Steady north efflux
C Roger, no contact
P negative

Comments

The Belgian skies south-east of Brussels were densely populated during the night March 30-31, 1990. There were at least three, and maybe six bright lights with changing colors, observed by gendarmes from 23 h (local time) to 1 h. There were two F-16 hunting true or false echoes from 0 h to 1 h. Local time = GMT + 2.

But there was something else. At 0 h 28, the Semmerzake radar detected an object 2500 ft over the western part of the Brussels agglomeration, moving towards Liege (roughly speaking, towards east) at 450 knots. At 0 h 29, the Glons radar detected it also. From 0 h 29 to 0 h 33, both radars followed the craft, which was going in straight line towards Liege, increasing its speed and its altitude. The Semmerzake radar spotted it again 6000 ft over Liege at 0 h 35, speed 650 knots. The last point was some 12 miles east of Liege, altitude 12000 ft, at 0 h 36.

The Semmerzake radar is an array type radar. It is used for military air safety. Semmerzake is about 30 miles west of Brussels.

Glons CRC is a part of NADGE (NATO Air Defense Ground Environment).

There are about 80 NADGE CRC in Europe (including Turkey). Its missions are: 1. detect and follow every flight in the Belgian air space, 2. identify friend or foe, 3. if foe, intercept and/or destroy according to the alert status. The Glons radar is a multipurpose impulsion type radar. Glons is about 6 miles north of Liege. The distance Brussels-Liege is about 60 miles.

There is another radar at Bertem, for civilian traffic. The craft passed 5 miles south of Bertem at 0 h 30. The Bertem radar did not see anything.

What was this?

1) Civilian traffic? I don't know anything in aviation, but I believed that airliners did not go that fast when they were that low. Also, I thought that each civilian flight was known to the military.

2) A little private jet? Maybe, if all this story is a hoax (it should be a really big hoax, with airship(s), jet(s) and probably ULMs). And are there private jets which go that fast?

3) A military plane? It was not one of the two F-16, which, instead of moving in straight line, were describing complicated loops (I have the plot of all the trajectories). In my opinion, it looks more like a stealth aircraft (F-117 A or TR-3 A), because it was seen only from certain angles. The Bertem radar, which was the best located, never saw it. But apparently the Belgian military did not know of this plane, at least in May 1990, when

the Belgian Air Force released its first report about the incidents:

00 h 32: The Glons and Semmerzake radars have a contact at 110 / 6 NM
away Beauvechain, which heads for Bierset at 7000 feet and high speed.
The registered speeds go from 478 to 690 knots. The contact is lost
above Bierset. The Maastricht radar control center has had no contact
with this UFO.

Summary

29 November 1989: Massive sightings by at least 30 groups of witnesses between Liege and Eupen.

30 March 1990: Belgian Air Force scrambles two F-16s in response to ground sightings; multiple radar locks obtained by both ground and airborne radar.

22 June 1990: Belgian Air Force releases detailed information on the

30 March 1990 sightings, including radar recordings.

Conclusion: 'Unable to identify either the nature or the origin of the phenomena.' Balloons, ultralights, remotely-piloted vehicles, conventional aircraft, stealth aircraft, laser projections, and mirages all ruled out.

6 November 1990: Various sightings all over Europe. Paris Air Traffic Control sees the Triangle but nothing shows on radar. Others report triangles, balls, lights, smokes, etc. Locations include France, Belgium, Germany, Switzerland, and Italy.

<u>Radar Contacts</u>

Observations: Radar Contacts F-16 interception 30/31 March 1990. The radar contacts of one F-16 with the so-called "UFOs" have been registered on a video record. One lasted for 46 seconds. Two F-16 were involved. One of the F-16 had 13 registered contacts; the other one had also contacts, but they were not registered because the pilot did not push the right switch. The contacts can be divided into 5 groups, separated by periods without contact.

contact number	lasting (seconds)	beginning at
1	2.3	00 h 13 March 31,1990 (March 30 22 h 13 GMT)
2	3.4	
3	19.9	00 h 15
4	27.5	00 h 29
5	8.0	
6	11.4	
7	9.3	
8	< 0.1	
9	45.9	00 h 39
10	16.2	
11	11.4	
12	9.5	
13	11.2	00 h 46

All positions and speeds are available on the video record. Speeds of vertical displacement seem to be sometimes above 1000 knots. Measured speeds are digitally displayed with only three numbers, so 200 knots could be 1200 or 2200 knots. Knowing that, positions and digits are dubious).

Here is the transcript of one of the contacts:

Seconds after	Heading	Speed	Altitude

lock-on	(degrees)		(knots)	(feet)	
00	200		150	7000	
01	200		150	7000	
02	200		150	7000	
03	200		150	7000	
04	sharp 200	acceleration	150	6000	
05	turn 270	= 22g	560	6000	
06	270		560	6000	
07	270		570	6000	
08	270		560	7000	
09	270		550	7000	
10	210		560	9000	
11	210		570	10000	
12	210		560	11000	
13	210		570	10000	
14	270		770	7000	
15	270		770	6000	
16	270		780	6000	
17	270		790	5000	
18	290		1010	4000	
19	290		1000	3000	
20	290		990	2000	
21	290		990	1000	
22	300		990	0000	
22.5	300		980	0000	Break lock

When you see for example altitude 5000, this means between 4500 and 5500. So 0000 means between 0 and 500. 0 is sea level; mean ground altitude in this area is about 200 feet (therefore 0000 means in fact between 200 and 500).

Appendix B: What Happens When We Spin Up the Negative Mass Ring?

There are limits to how fast a physical object can be spun. We know our ring will have to reach speeds 5,000,000 revolutions per second. Can it get there before disintegrating?

Ultracentrifuges can produce accelerations of about one million g. At higher rates of speed they disintegrate because of the centrifugal forces induced by the spin. The ultracentrifuges are about 20 cm in diameter, or a radius of .1 meters. How fast can they spin?

Radial acceleration is given by $\mathbf{a = \omega^2 r}$. The radial frequency $\boldsymbol{\omega}$ is 2π times the frequency of rotation **f.** Therefore

$$\mathbf{a = (2\pi f)^2 r}$$

Then the one million g's with $\mathbf{1\ g = 10\ m/sec^2}$ would be $\mathbf{1{,}000{,}000 \times 10\ m/sec^2 = 10{,}000{,}000\ m/sec^2}$

$$\mathbf{10{,}000{,}000\ m/sec^2 = 40 \times (.1) \times f^2}$$
$$\mathbf{= 4 \times f^2}$$
$$\mathbf{2.5(10)^6 = f^2}$$

or

$$\mathbf{1{,}600 = f}$$
$$\mathbf{f = 1{,}600\ revolutions/sec}$$
$$\mathbf{= 96{,}000\ revolutions/min.}$$

so ultracentrifuges can do about **100,000** revolutions per minute.

What about the our negative mass ring, which is much larger? For a **5 meter radius** ring the one million g's would be mean A rotation frequency of

$$\mathbf{1{,}000{,}000 \times 10 = 40 \times 5 \times f^2}$$
$$\mathbf{= 200 \times f^2}$$

$$5(10)^4 = f^2$$

or

$$f = 220 \text{ revolutions/sec.}$$
$$= 13{,}000 \text{ rpm.}$$

At which speed the ring might break up.

Initially the negative gravitational fields of the negative mass ring are so weak that they are suppressed by the positive gravity of the universe and confined to be within the negative mass. The negative mass at this point is subject to inertia.

But the inertia it experiences will be "perverse", in the opposite direction from normal inertia, in the opposite direction to the applied force.

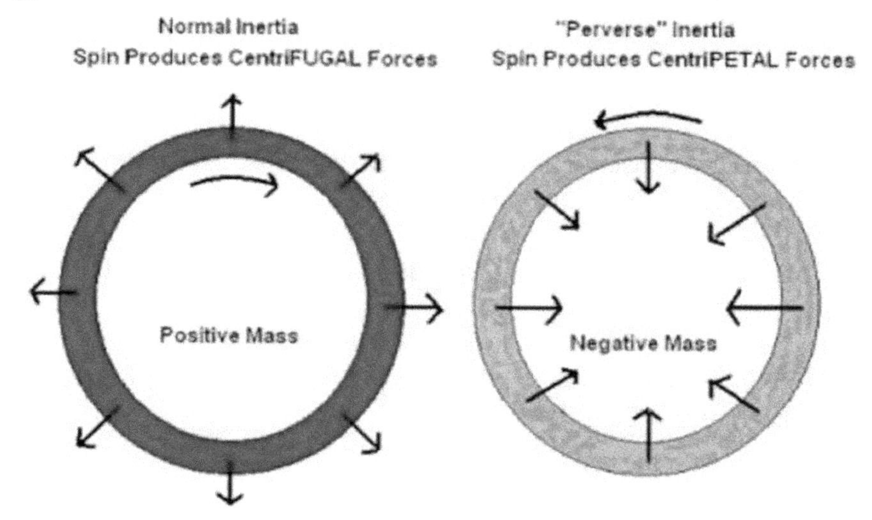

Clockwise electromagnetic spin force applied to ring.

The radial acceleration of the rotation will not create a CENTRIFUGAL force tending to tear the ring apart. It will generate a CENTRIPETAL force tending to compress the ring.

But it turns out that the resistance to compression of substances, for example metals, is not much greater for compression than it is for tension. Therefore the limit of 1 million g's still obtains, and the rotation limit for the ring would still be about 13,000 revolutions per minute.

But long before that happens, something else sets in. The rotation of the negative mass ring begins to generate a negative vector potential **A(-)**. A toroidal **A(-)** field now grows around the ring. This repels the positive **A(+)** potential of the universe which is responsible for inertia.

Suppose the saucer is stationary. We found in considering EM effects that they extended **about 1.5 miles**. That distance converted to a speed of, **v = .0006** m/sec.

Theoretically **A(+)** is zero when the UFO is stationary. But the EM data could allow for a "residual" background **A(+)**. The above speed times **1/G** indicates a possible "background" level of **A(+)**

$$\mathbf{A(+) = v/G}$$
$$\mathbf{= .0006 \times 1.5(10)^{10}}$$
$$\mathbf{= 9(10)^{6}}$$

that might cause inertia. The inertia would have to be overcome by the same amount of **A(-).**

As the ring speeds up the toroidal **A(-)** field will grow. It is given by

$$\mathbf{A(-) = \pi\kappa M f a^2/r^2}$$

The increasing **A(-)** toroid encompasses the ring at a frequency when **r = a = 5.**

$$\mathbf{9(10)^{6} = (3.14)4(10)^{4} \times f}$$
$$\mathbf{9(10)^{6} = 1.25(10)^{5} \times f}$$

or

$$\mathbf{800\ rps = f.}$$
$$\mathbf{4{,}800\ rpm = f}$$

This is about a third of the rotation speed where the 1 million g limit sets in.

Granted we have used inaccurate approximations to do the calculations. But it would seem, that the rotation will generate an inertia free region encompassing the ring well before the compression break up limit is reached.

From then on inertia is no longer in the picture for the spin. The ring is essentially inertially massless. Only residual air in the ring chamber and inductive forces hinder the spin. The lights without beams, which we call plasma ports, emit about 10 kilowatts. We presume this is the energy used to keep three rings spinning. That is about the energy of 13.4 horsepower, or the output of a 200 cc motor.

Glossary of Symbols

φ gravitation scalar potential of finite objec
Φ background scalar potential of universe
$\nabla\varphi$ gradient of scalar gravitational potential
ρ mass density
κ coefficient for calculations in negative gravitational bubble
∇v discontinuous change in velocity
gravitational dipole moment
λ linear mass density
a acceleration
A gravitational vector potential
A(-) negative gravitational vector potential
A(+) positive gravitational vector potential
$\partial A / \partial t$ change of vector potential with time
c speed of light
E gravitational field
EI induced electric field
e/m_i electron charge to mass ratio
F frequency
G gravitational constant
g acceleration due to gravity
I_m mass current
F_i force of inertia
F_g force of gravity
k spring constant
K.E. kinetic energy
L length of pendulum
m_i inertial mass
m_g gravitational mass
$m_i v$ momentum
N Newton, unit of force
T period or oscillation
U total energy of oscillation
v velocity

About the Author

The author is theoretical physicist who did his graduate work in elementary particle theory at MIT in the 1960s. He became interested in the subject of UFOs only three years ago.

CPSIA information can be obtained
at www.ICGtesting.com
Printed in the USA
BVHW060021301222
655301BV00009B/209